SCENES FROM PROVINCIAL LIFE

SCENES FROM PROVINCIAL LIFE

Knightly Families in Sussex 1280–1400

NIGEL SAUL

CLARENDON PRESS · OXFORD
1986

Oxford University Press, Walton Street, Oxford OX2 6DP
Oxford New York Toronto
Delhi Bombay Columbo Madras Karachi
Petaling Jaya Singapore Hong Kong Tokyo
Nairobi Dar es Salaam Cape Town
Melbourne Auckland
and associated companies in
Beirut Berlin Ibadan Nicosia

Oxford is a trade mark of Oxford University Press

Published in the United States
by Oxford University Press, New York

British Library Cataloguing in Publication Data
Saul, Nigel
Scenes from provincial life: knightly families in Sussex 1280–1400.
1. Sussex—Gentry—History
I. Title
305.5'232'094225 HT657
ISBN 0-19-820077-3

Library of Congress Cataloging in Publication Data
Saul, Nigel.
Scenes from provincial life.
Bibliography: p.
Includes index.
1. Sussex—Gentry—Social life and customs.
2. Knights and knighthood—England—Sussex. 3. England—Social life and customs—Medieval period, 1066–1485.
4. Sussex—History. I. Title.
HT657.S39 1986 305.5'232'094225 86-12631
ISBN 0-19-820077-3

Typeset by Joshua Associates Limited
Printed in Great Britain
at the University Printing House, Oxford
by David Stanford
Printer to the University

FOR MY WIFE

PREFACE

THIS book is not another study of a gentry-led 'county community'. Nor is it a study of one man in his time, like Colin Richmond's inimitable *John Hopton* (Cambridge, 1981). It lies somewhere in between. It looks at three families—the Etchinghams, the Sackvilles, and the Waleyses—and, through them, at society in the corner of east Sussex where they lived.

These three families were richer than many, perhaps most, of their peers in Sussex; but, that apart, they constitute a reasonably representative cross-section of fourteenth-century landed society. Between them they can provide several magnate retainers, a royal household knight and even a Chancellor of the University of Oxford. However, it is less the typicality, or even the individuality, of these people that justifies the present study than the fact that, by the modest standards of the medieval gentry, their lives are comparatively well documented. To that extent they selected themselves.

The main source has been the collection of account rolls and charters which has come to rest in the Glynde Place archives in the East Sussex Record Office at Lewes. Their survival is not, as is so often the case, a consequence of continuity of ownership. Quite the opposite, in fact. The Glynde estate has changed hands several times in its history. On the first occasion there was trouble, and the difficulty that the new owners—the Morleys—had in establishing their title may explain why they held on to many documents that might otherwise have been discarded. Thanks to that accident, and to the patient work in more recent times of local archivists, we have both charters and account rolls from the manor of Glynde, which was the main seat of the Waleys family, and account rolls alone from the neighbouring manor of Beddingham, which was held in the middle ages by the Etchinghams and later acquired by the Waleyses' successors, the Morleys.

I have been able to supplement the sources in the Glynde collection by using the account rolls of the Sackville manors of Chalvington and Claverham, which are also deposited in the East Sussex Record Office, and those of Udimore, another Etchingham manor, in the Ray collection at Hastings Museum. This is therefore a study based principally on research in local, not national, archives—and rightly so, because the riches that lie in our local record offices have until recently received but little attention from historians of the medieval gentry. However, it goes without saying that I have also drawn upon the documentation in the Public Record Office to illuminate the interaction between the lives of these families and the workings of the central government.

The debts that I have incurred along the way are many. First there are those that I owe to others who have laboured in the vineyard of Sussex history long before me, and pre-eminently to the two antiquarian scholars without whose patient labours my own preparatory work would have had to be much greater than it has been—that is to say, in the eighteenth century William Hayley, who made a detailed description of the stained glass windows in Etchingham church before their destruction, and in the early years of this century Edwin Dunkin, who devoted a lifetime to sifting the public records for material relating to Sussex. His manuscript notebooks in the British Library are a mine of information for anyone interested in the history of this county.

My debts to friends and colleagues who have commented on early drafts of parts of this book again are many, and are acknowledged individually in footnotes. But there are two debts which ought to be acknowledged now: the first is to Barbara Harvey, who has several times discussed with me the problems facing medieval landowners, and the second is to Christopher and Margaret Whittick, whose knowledge of the record office at Lewes has greatly facilitated my researches and whose kind hospitality has made my visits to Sussex over the past five or six years a source of such lasting pleasure.

In conclusion, I take this opportunity to thank the British Academy, the small grants fund of Royal Holloway College, and the central research fund of the University of London for their generosity in making grants to support various stages of this work.

February 1986 Nigel Saul

CONTENTS

	List of Tables and Genealogies	x
	List of Plates	xi
	Abbreviations	xii
I	The Families	1
II	The Local Community	28
III	The Common Law	73
IV	The Estates and their Management	98
V	Etchingham Church	140
VI	The Families' Life-Style	161
	Bibliography	193
	Index	201

TABLES AND GENEALOGIES

TABLES

I The Litigation of the Etchinghams, 1300–1412 79

II The Litigation of the Waleys Family, 1300–1418 79

III The Litigation of the Sackvilles, 1300–1410 79

IV The Manor of Beddingham 117

V Cash Wages of Demesne Staff at Glynde 122

VI Cash Wages of Demesne Staff at Beddingham 123

VII Cash Wages of Demesne Staff at Udimore 124

VIII Cash Wages of Demesne Staff at Chalvington and Claverham 124

IX Udimore: Sales of Pasture Rights 137

GENEALOGIES

Etchingham 2

Sackville 8

Waleys 15

Pashley 87

DIAGRAMS

Etchingham Church—Plan of the Stained Glass Windows 150

LIST OF PLATES

Between pages 164 and 165

I. The brass of Sir William de Etchingham V (died 1389) in Etchingham church (from a rubbing by the author, reproduced by kind permission of the Rector and Churchwardens of Etchingham)

II. Etchingham church: the chancel looking west, showing the stalls and, on the floor, the brasses of Sir William V and Sir William VI (reproduced by kind permission of the Managing Editor of *Sussex Life*)

ABBREVIATIONS

Arch. Cant.	*Archaeologia Cantiana*
BL	British Library
Cal. Robertsbridge Charters	*Calendar of Charters and Documents relating to the Abbey of Robertsbridge . . . Preserved at Penshurst among the Muniments of Lord de Lisle and Dudley* (London, 1873)
CCR	*Calendar of Close Rolls*
C.CH.R	*Calendar of Charter Rolls*
CFR	*Calendar of Fine Rolls*
CIPM	*Calender of Inquisitions post mortem*
CPR	*Calender of Patent Rolls*
EHR	*English Historical Review*
Ec.H.R	*Economic History Review*
ESRO	East Sussex Record Office
Feud. Aids	*Feudal Aids*
GEC	Cokayne, G. E., *The Complete Peerage*, ed. V. Gibbs and others (12 vols. in 13, London, 1910–59)
Jnl. Eccles. Hist.	*Journal of Ecclesiastical History*
Lewes Cartulary	*Chartulary of the Priory of St Pancras, Lewes* (SRS xxxviii and xl, 1932 and 1934)
Moor, *Knights*	C. Moor, *Knights of Edward I* (5 vols., Harleian Soc. 80–4, 1929–32)
PRO	Public Record Office
SAC	*Sussex Archaeological Collections*
SRS	Sussex Record Society
Three Earliest Subsidies	*Three Earliest Subsidies for the County of Sussex*, ed. W. Hudson (SRS x, 1910)
TRHS	*Transactions of the Royal Historical Society*
VCH	*Victoria History of the Counties of England*

All documents quoted are in the East Sussex Record Office at Lewes unless otherwise stated.

I

THE FAMILIES

On Saturday 9 May 1366 a group of Sussex gentlemen met at Sir William de Etchingham's manor-house at Etchingham to assist one of their neighbours in making a settlement of his property.[1] Among those whom the host welcomed that day were two fellow knights, Sir Andrew Sackville and Sir John Waleys, two esquires, Robert de Ore and William de Batsford, and a local sergeant-at-law, William Tauk.[2] They had assembled to witness a quitclaim to Sir Robert de Pashley from his feoffees of his manor of The Moat, at Leigh-in-Iden near Rye. Sir William, Sir Andrew, and Sir John were the obvious men for him to approach for the favour of attestation. They were neighbours or near neighbours, and friends rather than rivals. They and their forebears had wined and dined with each other, hunted with each other, and seen active service together.[3] It is purely by accident that the estate documents of three of the families should have come to rest side by side in the same location at Lewes. But it is also both fitting and appropriate. For the ample evidence that they formed, if not a 'community', then a circle of their own in this corner of east Sussex lends a measure of logic to the idea of treating them as a group for the purposes of this study. But groups are made up of individuals, and individuals need to be introduced. Courtesy and good manners require that we make these introductions our first task.

1. THE ETCHINGHAMS

Arms: Azure fretty argent

Sir William de Etchingham, the host, was the current head of one of the oldest and most distinguished knightly lineages in Sussex. They

[1] *CCR 1364–8*, p. 289.

[2] For Tauk see J. B. Post, 'The Tauke Family in the Fourteenth and Fifteenth Centuries', *SAC* 111 (1973), 93–107. William de Holmestede, Etchingham's steward, who was also present, may have been a laywer too. He was described as an 'apprenticus de leye' in 1379 (PRO, E179/189/41).

[3] The evidence for this statement will accumulate in the course of the book, particularly in Ch. II.

The Etchingham Family

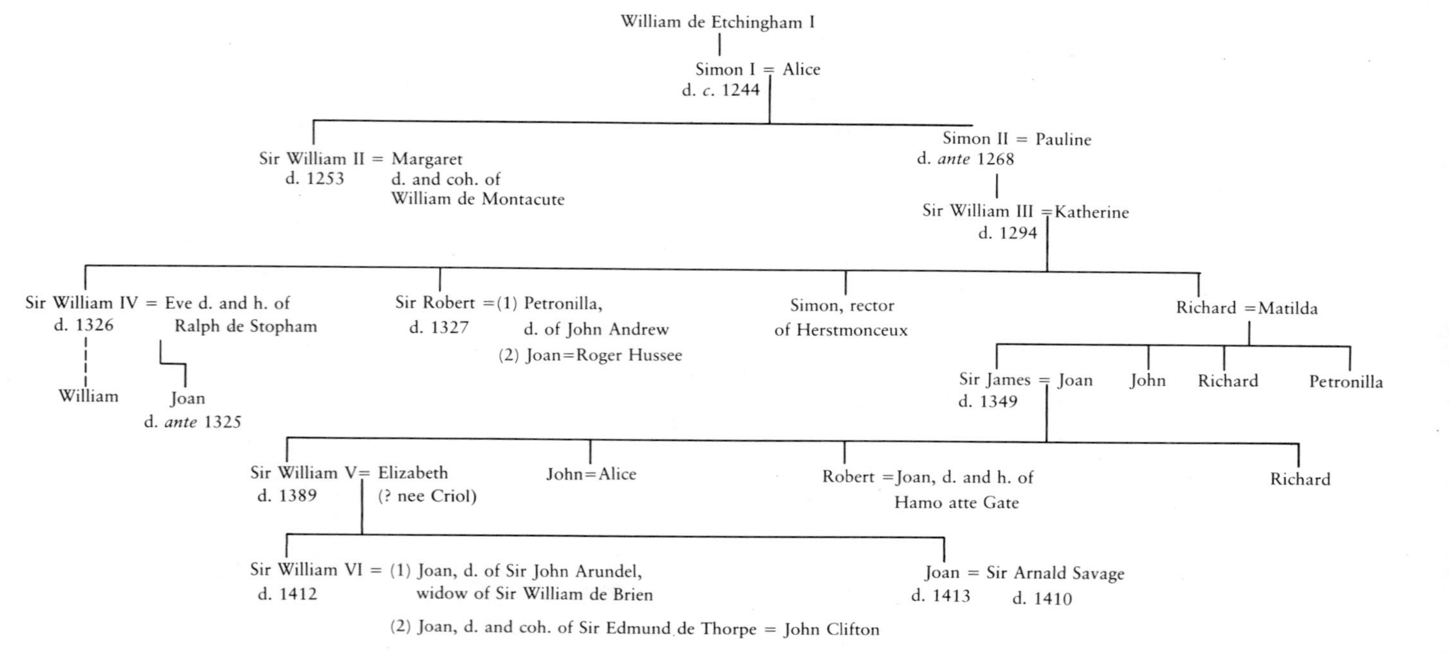

took their name from the Wealden manor of Etchingham, which was their principal seat and residence. But they also held manors at Salehurst, Mountfield, Beddingham, Peakdean, and Udimore, all in Sussex, Brenzett, Lullingstone, and Hempsted in Kent, and Padbury in Buckinghamshire.[4] The income which Sir William drew from this string of properties made him probably the richest of the four landowners meeting that day in May 1366.[5] But three-quarters of a century earlier his great-uncle, another Sir William, had almost certainly been richer still. His marriage to Eve, daughter and co-heiress of Ralph de Stopham, had brought him the manors of Stopham, Yapton, and Linch in Sussex, and those of Bryanston and Bradford Bryan in Dorset.[6] By 1311 he was wealthy enough to be summoned to parliament as a lord. On two occasions in the past he had been elected to the Commons as a knight of the shire for Sussex. But for the remainder of his career he was chosen to sit with the earls and barons. He was the only one of his line to be so privileged, however. The honour had not yet become hereditary. It depended in varying degrees on royal favour, personal ability, and the possession of a certain amount of landed wealth. Sir Robert, Sir William IV's brother and successor, certainly possessed royal favour and personal ability.[7] But the gains in landed endowment that his elder brother had registered by his marriage to the Stopham heiress did not last.

[4] There is a list of the family's manors in a grant of free warren made in 1295 (*C.Ch.R.1257–1300, p. 461*). For the Sussex manors see further *VCH Sussex*, ix. 173, 212, 235, and for Padbury, *VCH Bucks.* iv. 211–12. For the Kent manors, the descent of which is harder to trace, see 'Kent Fines, Edward II', ed. J. Greenstreet, *Arch. Cant.* xiii (1880), 295, and W. E. Ball, 'The Stained Glass Windows of Nettlestead Church', ibid. xxviii (1909), 231. In 1377 Sir William Etchingham V conveyed Padbury to feoffees who in their turn conveyed it two years later to Sir John Hawkwood, the famous *routier* (*VCH Bucks.* 212; *CCR 1377–81*, pp. 91–2). The same Sir William also disposed of Hempsted (Kent), in this case to Sir Robert Belknap, the Chief Justice of Common Pleas. Along with the other justices who answered Richard II's questions on the nature of the royal prerogative, Belknap forfeited his lands in 1388, and Sir William VI recovered Hempsted three years later (PRO SC8/270/13496; *CCR 1389–92*, pp. 383–4).

[5] The income from the manors of Beddingham and Udimore, for which series of account rolls survive, is discussed below, in Ch. IV. For Etchingham and Mountfield there are a few beadles' accounts from the 1380s which show a rental income of £43 due from the former and £17 from the latter. The Etchingham figure probably includes Salehurst as well (Hastings Museum JER/Box 3). I am grateful to Victoria Williams, the curator of Hastings Museum, for allowing the documents in her keeping to be deposited temporarily in the East Sussex Record Office at Lewes.

[6] *VCH Sussex*, iv. 65–6; J. Hutchins, *The History and Antiquities of the County of Dorset* (4 vols., Westminster, 1861), i. 248–51.

[7] For Robert see below, pp. 000. For the evolution of the parliamentary peerage see K. B. McFarlane, *The Nobility of Later Medieval England* (Oxford, 1973), pp. 268–78.

Bryanston Sir William had granted or sold to his friend Sir Alan Buxhill;[8] and Stopham, Linch, and Yapton were also to pass into other hands.[9] It was the manors in east Sussex which were to form the core of the inheritance for the next century and a half. They were not quite numerous enough to qualify Robert and his successors for a personal summons; but they were sufficient to put them amongst the richest of the county élite.

The earliest member of the family who can be identified with any certainty is Simon de 'Achingham' who was living in about 1150.[10] His father, Drew of Pevensey, was in turn the heir of Reinbert, the first Norman steward of the lords of Hastings Rape, the counts of Eu.[11] As a later member of the family was to lay claim to the stewardship, it seems that Reinbert may be regarded as the founder of the Etchingham family.[12] The fortunes of these early Etchinghams were closely bound up with those of the counts of Eu, whose charters they are often found attesting. But in the thirteenth century, as the counts withdrew from the scene, they began to emerge as important figures in their own right—in the county, moreover, as well as in the rape. Simon (d. *c.* 1244) was the first of his line to hold the shrievalty.[13] And William III (d. 1294), his grandson, served at various times as a keeper of the peace, assessor of subsidies, and justice of oyer and terminer.[14] It was his son, William IV, who received the summons to parliament in 1311. As we have seen, his income from land more than justified it. So too did his record as a *miles strenuus*. He was summoned to perform military service in Flanders in 1297 and Scotland in 1298, and is known to have responded in the latter year, when he was present at the battle of Falkirk, and again in 1299 and 1301.[15]

William's marriage to the heiress Eve de Stopham brought him additional manors but no sons to whom he might transmit them. His only children were a daughter Joan and a bastard son who was later foisted off on to Robertsbridge Abbey.[16] When he died in 1326 he was succeeded by his brother Robert, a careerist knight, who had done

[8] *CPR 1313–17*, p. 161. [9] *VCH Sussex*, iv. 65–6.

[10] Ibid. ix. 212. Spencer Hall, *Echyngham of Echyngham* (1850) purports to be the history of the family, but is too unreliable to be of any value.

[11] E. Searle, *Lordship and Community: Battle Abbey and its Banlieu, 1066–1538* (Toronto, 1974), p. 210. [12] Hall, p. 2.

[13] *List of Sheriffs*, pp. 135–6.

[14] *CPR 1272–81*, p. 69; *1281–92*, pp. 105, 265, 404, 453, 514.

[15] For summonses see Moor, i. 299–300. For service actually performed see PRO C67/14 mm. 4, 16, and H. Gough, *Scotland in 1298* (Paisley, 1888), p. 213. The last lists his company at Falkirk. [16] BL Add. MS 28,550, fol. 5.

well for himself in the service of Edward II and the Despensers.[17] Within eighteen months he too had died, like his brother without male issue, and the estates passed, though not without some opposition from Robert's widow, to the next brother, Simon, the rector of Herstmonceux—a quarrelsome man as we shall see.[18] He in his turn was succeeded, some time in the late-1330s, by his nephew James, son of Richard, the youngest of the four brothers.[19] James was a less assertive character than his three uncles had been. He shouldered such administrative duties as came his way, although these never included the shrievalty or election to parliament.[20] But he fought shy of assuming knighthood. He obtained a respite for three years in 1342, and renewed it in 1346 shortly before setting out for France.[21] But by then there was no point in further delay. Remaining an esquire merely condemned him to taking half the rate of pay he would have enjoyed as a knight. Self-interest took over; and by September he had been dubbed.[22]

Three years later James died, leaving as his heir a son, William V, who was still a minor.[23] Before setting out on active service, however, he had taken the wise precaution of granting his estates to feoffees, on the understanding that in the event of his death they would enfeoff his son when he came of age.[24] The two feoffees he chose were John de Ore, a long-standing family friend, and his brother Master John de Etchingham, an Oxford academic. John was someone altogether more substantial than the typical tonsured younger son. He must have gone up to Oxford in the early-1340s, because the degree of MA, which he gained in 1348, was only awarded after seven years' study. After another nine years he was awarded the baccalaureate of theology and finally in 1365 the doctorate in the same faculty. The work involved in supplicating for these higher degrees was long and arduous; but John had livings to support him. In March 1357 he was collated to the deanery of South Malling, near Lewes; and in 1362 he was granted a

[17] See below, pp. 51–5.

[18] See below, pp. 94–7.

[19] The last reference to Simon is in the Herstmonceux court roll for 9 Dec. 1337 (Harvard Law School Library, no. 69). I owe this reference to Christopher Whittick. All these relationships are made clear in the lists of essoins in PRO, JUST 1/938/1 m. 55.

[20] The nature of his work in local administration can be picked up from *CPR 1338–40*, pp. 362, 559; *1343–5*, *pp.* 79, 416, 515; *1348–50*, p. 78.

[21] *CPR 1340–3*, p. 438; *CPR 1346–9*, p. 85.

[22] *CCR 1346–9*, p. 153.

[23] He died, according to the inquisition *post mortem*, on 22 Aug. 1349 (*CIPM* ix, no. 601) and not in 1350 as the *VCH* says (*VCH Sussex*, ix. 212). His son was said to be aged '16 and more'.

[24] PRO, JUST 1/941A m. 32.

prebend in St Paul's.[25] The income from these benefices sustained him in residence at Oxford in the 1360s—in 1363 he served a term of office as Chancellor. But it also enabled him after that to live the life of a country gentleman in Sussex, given over to litigating with his neighbours and hunting the abbot of Battle's game.[26] He was the sort of clerk who would have had much in common with Chaucer's Monk.

Though it was John who took holy orders, it was his nephew William, James's son, who was to prove himself the more devoted son of the Church: or, at least, the more generous benefactor. For, as the inscription on his brass tells us, it was he who rebuilt Etchingham church.[27] Begun in the early-1360s, it was a project which eventually absorbed him to the exclusion of the more mundane duties that fell to men of his rank. Office-holding attracted him little:[28] he was never appointed sheriff or elected to parliament. In the later fourteenth century it was his younger brother Robert, of Greater Dixter, who played the more active role in county affairs.[29] But that is not to say that William was a self-effacing man. Far from it. He made the parish church the focus of family pride. By filling its windows with the coats-of-arms of the Etchinghams and their neighbours he made it a complex symbol which bore witness to the family's position in the world; and by choosing burial there he ensured that it would be his

[25] A. B. Emden, *A Biographical Register of the University of Oxford to A. D. 1500* (Oxford, 3 vols. 1957–9), i, 236.

[26] Pleas of trespass were brought against John in 1355 (PRO, KB27/380 m. 73d) and 1371 (KB27/440 m. 24). Before 1352 he had received Thomas de Pashley, an outlaw (PRO, JUST 1/941A m. 36). In 1374 an assize of novel disseisin was brought against him by the prior of Christ Church, Canterbury (PRO, JUST 1/1484 m. 6d). Before 1379 his property at Brasted (Kent) had been attacked (PRO, KB27/474 m. 9d). He frequently acted as a feoffee (for example, JUST 1/941A m. 6d and *Rotulorum Originalium in Curia Scaccarii Abbreviatio*, ii (1810), p. 286). For a visit he paid to Udimore to hunt in the park see the Udimore account for 1362–3 (Hastings Museum, JER/Box 3).

[27] Etchingham church is discussed in detail in Ch. V.

[28] His experience of the consequences of being appointed a collector of the notorious poll tax granted in November 1380 was enough to dissuade him from doing that ever again. By the time he was appointed once more, in 1382, he had obtained an exemption (*CFR* 1377–83, pp. 225, 340; and see below, pp. 81–2).

[29] Robert was sheriff in 1390/1. He married Joan, daughter and heiress of Hamo atte Gate of Dixter (in Northiam) (L. F. Salzman, 'Descent of the Manor of Dixter', *SAC* 52 (1909), 153). It is tempting to associate the building of the present house at Dixter with Robert's occupancy since he was of more distinguished lineage than his wife. The architectural evidence, however, favours a later date. The arms of Etchingham are to be observed on the faces of two of the hammer-beams in the hall (J. E. Ray, 'Dixter, Northiam: a Fifteenth Century Timber Manor House', *SAC* 52 (1909), 143, opposite which there are illustrations).

name and not that of his brother that would be remembered by posterity.

His successor was his son, yet another William. With the exception of the fourth William he is the only member of the family whose wives can be identified by surname as well as by Christian name.[30] This information is of more than purely antiquarian worth: for, if a man's standing can be measured by that of the families into which he married, then this William indeed stood high in the estimation of his peers. His first wife was Joan, the eldest daughter of Sir John Arundel and widow of Sir William de Brien of Seal (Kent), and his second Joan, daughter and coheir of Sir Edmund de Thorpe of Ashwellthorpe (Norfolk).[31] Moreover, William's sister, another Joan, was married to Sir Arnald Savage of Bobbing (Kent), a future Speaker of the House of Commons.[32] These families were his equals, not his superiors. But some can be more equal than others, these families among them.

II. THE SACKVILLES

Arms: Quarterly, or and gules, a bend vair

Sir Andrew Sackville was of no less distinguished pedigree than his

[30] One other possible exception ought to be mentioned. Sir William V's wife may have been a member of the distinguished Kentish house of Crioll. This possibility is suggested by the bequest made in his will, dated 1379, by Sir Nicholas Crioll 'to Elizabeth Etchingham, 20 marks', and by his appointment of Sir William Etchingham and Robert Etchingham as two of his executors (*Testamenta Vetusta*, ed. N. H. Nicolas (2 vols., London, 1826), i. 103). Sir William's wife's Christian name is known to have been Isabella (i.e. Elizabeth) (PRO, CP40/532 m. 111d). This connection would explain the presence of the Crioll arms on the brass in Etchingham church of William V's son and grandson, illustrated and discussed by C. E. D. Davidson-Houston, 'Sussex Monumental Brasses', *SAC* 77 (1936), 168–71.

[31] Sir William VI, his first wife and their son Sir Thomas are commemorated by the triple canopied brass mentioned above and illustrated in *SAC* 77. The parents of the second Joan are commemorated by a fine tomb in Ashwellthorpe church, Norfolk (N. Pevsner, *North-West and South Norfolk* (Harmondsworth, 1962), p. 77 and plate 42).

[32] The relationship is made clear in the will of her son, another Sir Arnald (d. 1420), printed in *The Register of Henry Chichele, Archbishop of Canterbury, 1414—1443* , ii, ed. E. F. Jacob (Oxford, 1938), pp. 205–6. He set aside 20 marks for a brass to be laid in Bobbing church (Kent) to the memory of his parents. The female figure, he said, was to bear the arms of Etchingham on the kirtle and those of Savage on the mantle. In fact the figure as executed had neither: it was left plain. Both this brass and that of the son are illustrated in W. D. Belcher, *Kentish Brasses*, i (London, 1888), p. 15. Joan, incidentally, seems to have left her Sussex connections behind her when she married and settled in Kent. Her will, printed by G. O. Bellewes, 'The last Savages of Bobbing', *Arch. Cant.* xxix (1911), 167, makes no mention of her Etchingham relatives. William Etchingham VI and his uncle Robert, of Dixter, had fought alongside the elder Arnald Savage in the naval expedition led by the earl of Arundel in 1387 (PRO, E101/40/33).

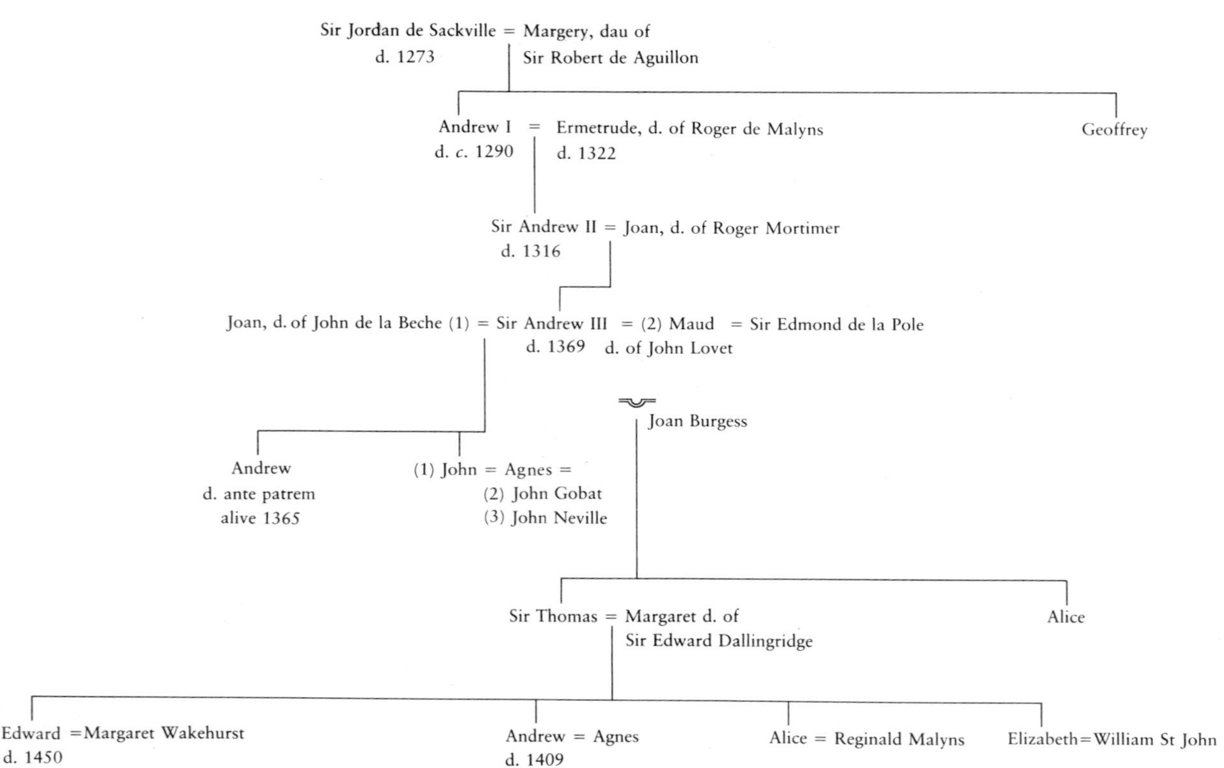
The Sackville Family
Sir Jordan de Sackville = Margery, dau of
d. 1273
Sir Robert de Aguillon
Andrew I = Ermetrude, d. of Roger de Malyns
d. *c.* 1290
d. 1322
Geoffrey
Sir Andrew II = Joan, d. of Roger Mortimer
d. 1316
Joan, d. of John de la Beche (1) = Sir Andrew III = (2) Maud = Sir Edmond de la Pole
d. 1369
d. of John Lovet
Joan Burgess
Andrew
d. ante patrem
alive 1365
(1) John = Agnes =
(2) John Gobat
(3) John Neville
Sir Thomas = Margaret d. of
Sir Edward Dallingridge
Alice
Edward = Margaret Wakehurst
d. 1450
Andrew = Agnes
d. 1409
Alice = Reginald Malyns
Elizabeth = William St John

host. He claimed descent from Herbrand, the lord of Sacqueville in Normandy, who had come over with the Conqueror and had been rewarded with the manor of Fawley (Bucks). There the senior branch of the family was to reside for the rest of the middle ages. But a junior branch was established in Sussex shortly after 1200, when Jordan de Sackville acquired the manor of Buckhurst by marriage to the heiress Ela de Dene.[33] Many centuries later the Sackvilles of Buckhurst were to eclipse their Fawley kinsmen in fame and wealth by rising to the commanding heights of the peerage as dukes of Dorset. But in the middle ages they remained a family of knightly rank, resident chiefly at Buckhurst, but not as closely identified with the county of Sussex as the Etchinghams were. The reason was that their estates were more scattered. They held the manors of Buckhurst itself (in the parish of Withyham), Chalvington, Bowley (in Hailsham), and Amberstone, all in Sussex, Debenham in Suffolk, Bures Mount and Bergholt in Essex, and Emmington in Oxfordshire.[34] Given that these three blocks of manors lay at a good distance from each other, their owners needed to be constantly on the move between them to ensure that their interests were neither threatened nor neglected. And this they seem to have realized. They were to be found, for example, at Emmington in 1306, when a son was born, and at Bures in 1342, when Sir Andrew witnessed a charter.[35] Two marriages into the Malyns family of Chinnor further attest the importance they attached to strengthening their Oxfordshire connections.[36]

Like the Etchinghams, the Sackvilles were therefore a well-endowed family. If they were not yet poised to enter the peerage, they nevertheless had no difficulty in sustaining the rank of knighthood and the style of life that went with it. Their level of income, however, is hard to estimate accurately because their manors varied widely in size and therefore in value. Some, like Chalvington (Sussex) and Emmington

[33] For Ela see L. F. Salzman, 'Some Sussex Domesday Tenants, II: the Family of Dene', *SAC* 58 (1916), 177. A full and accurate account of the medieval Sackvilles is to be found in C. J. Phillips, *A History of the Sackville Family*, i (London, 1929).

[34] For a list of the properties held by Sir Andrew II at his death in 1316 see *CIPM* v, no. 612. There is another list in the grant of free warren to Sir Andrew III (*CPR 1345–8*, p. 529). In 1340 Andrew III acquired the manor of Southey on the edge of Pevensey levels (*CCR 1339–41*, p. 453). For Debenham, where the Sackvilles had a market (*CCR 1313–18*, p. 350) see W. A. Copinger, *The Manors of Suffolk* (7 vols. Manchester, 1905–11), pp. 137–8. For Bures see P. Morant, *History & Antiquities of the County of Essex* (2 vols., London, 1763–8), ii. 224–5.

[35] *CCR 1341–3*, p. 525; Phillips, pp. 81–2.

[36] Phillips, p. 70.

(Oxon.), were coincident with villages. Others, like Debenham and Bergholt, in East Anglia, were not. According to the extents taken on the death of Sir Andrew II in 1316 Buckhurst was worth £24, Chalvington and Bowley £17 together, Emmington £11. 10*s*. 0*d*. and Debenham and the other lands in eastern England £5. 10*s*. 0*d*., giving a total of £58.[37] But figures in inquisitions *post mortem* are estimates not of the annual value of the manors to their lord but of the rent that could be realized from leasing out.[38] A very different impression is given by the view of account compiled for the Sackville feoffees in about 1370 shortly after the death of Sir Andrew III.[39] This is not a comprehensive document; indeed, it is more in the nature of an inventory. But it does give a clear idea of the level of underassessment recorded in the earlier inquisitions. The feoffees received £5. 12*s*. 0*d*. from the farm of Bergholt for Michaelmas term 1369, suggesting an annual total of the order of £20, £6 from the farm of Bures for the same term, suggesting an annual total of £24, and sales of grain from Emmington to the value of £33. On the basis of this evidence, a figure of at least £50–60 would seem possible for the properties in Essex and Suffolk and of perhaps £100–150 for Emmington and the Sussex manors. An overall value of, say, £200 or more would certainly be a better guide to the Sackvilles' income from land than the £58 recorded in the inquisition of 1316. But it should not be regarded as any more than an informed guess.

The medieval Sackvilles were all in their different ways active men, whose public and private lives offer much of interest to the student of family fortunes. Jordan de Sackville (d. 1273) was a supporter of Simon de Montfort, and was captured at the battle of Evesham on 4 August 1265.[40] His lands were taken into custody, and it is likely that the cost of redeeming them was to place a strain on the family finances for at least a generation. In May 1279 Jordan's son Andrew requested, and was granted, a postponement until the next parliament of repayment of the debts for which he was liable at the Exchequer.[41] And only five months earlier he had acknowledged that he owed a debt of £10 to Henry de Lenn, interestingly a clerk of Bishop Burnell, the king's

[37] *CIPM*, v, no. 612.

[38] E. A. Kosminsky, *Studies in the Agrarian History of England in the Thirteenth Century* (Oxford, 1956), p. 48.

[39] SAS/CH 258. The contents of this document are discussed in more detail below, pp. 171–3.

[40] Phillips, p. 62.

[41] *CCR 1272–9*, p. 258.

Chancellor and a well known dealer in encumbered lands.[42] This Andrew, however, was destined from childhood to be a king's man. A minor when his father died, he was brought up as a royal ward and married to Ermetrude, a lady-in-waiting of Queen Eleanor.[43] He is found in the queen's company in 1280–1.[44] In 1282 he experienced what was probably his first taste of active service in Edward's second Welsh war, and four years later he accompanied the king to Gascony.[45] But he did not live to take part in the long period of warfare in Scotland that opened in 1296. It was his son Andrew II, who was to distinguish himself in that theatre, both before and after his formal dubbing, alongside 200 other tiros, at the Feast of Swans in 1306.[46] This Andrew's career was to be of no longer duration, however, than those of his father and grandfather. He died in 1316, before reaching forty, and left a son under age and a widow exposed to the dangers of ravishment and abduction.

The son was made a ward of John de la Beche, a royal household knight.[47] His mother, it seems, was assigned the manor of Chalvington in dower; certainly it was there that she took up residence, and there that her several enemies found her. First on the scene were John Arnald, a London goldsmith, and one Roger de Blakehou of Tewkesbury (Gloucs.).[48] To judge by the Londoner's occupation, they had probably come in force to collect debts that Andrew still owed at the time of his death. However, the more serious challenge was to come shortly afterwards from a gang headed by, or at least including, Sir John Maufe and his accomplice Bartholomew Aubry.[49] Maufe must have made off with Joan—or tried to make off with her—because in July 1319 at the behest of his patron Bartholomew Badlesmere he obtained a pardon for the offence from the king.[50] But Aubry too was reputed to have ravished her, indeed to have pursued her as far as St Omer, near Calais, on the pretext that she was his wife.[51] Whether he was acting on his own initiative or on Maufe's we are not told.

[42] Ibid., p. 554. According to Sir Maurice Powicke, Bishop Burnell 'built up a wide-spread complex of landed property by purchase, exchange, the conversion of loans and other ways . . .' (*The Thirteenth Century* (Oxford, 2nd edn., 1962), pp. 338–9).

[43] *CCR 1272–9*, p. 192.

[44] *CCh.R 1257–1300*, pp. 234, 256.

[45] Moor, *Knights*, iv. 198–9; *CPR 1281–92*, p. 239.

[46] Phillips, p. 74; PRO, C67/14 mm. 6, 8; C67/15 m. 13.

[47] *CFR 1307–19*, p. 303.

[48] PRO, KB27/238 Rex m. 19d.

[49] PRO, KB27/234 m. 130.

[50] PRO, KB27/238 Rex m. 5.

[51] *CPR 1317–21*, p. 328.

Charges of ravishment and abduction, of course, can sometimes conceal what in reality was an elopement, and by taking them at face value we may be in danger of following an entirely false lead. It may, for example, be of greater significance that further down the list of those involved in the attack appears the name of Thomas Fauconer, a former servant of Sir Nicholas de la Beche. If the presence of this erstwhile dependant is a sign that other members of the de la Beche clan were involved, then the object of the raid might simply have been to eject Joan from the dower lands which she had been assigned.

In the absence of any further entries on the plea rolls it may be assumed that thereafter Joan was left in peace. But the period of wardship could easily have proved disastrous to the family fortunes. That it did not must have been due in some measure to the character and personality of Joan's son, Andrew III. A successful soldier and active careerist, he was to make himself at various times the servant of no fewer than four noble families—the Mortimers, the Despensers, the Montagus, and the FitzAlans.[52] It is possible that he had been launched on his way by his mother, who was herself a Mortimer. But it is unlikely: he would not have seen much of her as he grew to manhood. At the age of ten he had passed, as we have seen, into the wardship of Sir John de la Beche, a royal household knight who would have been well placed to pick up such perquisites. The sum he offered was £120, half to be paid in November 1316, and half the following February.[53] It was a bargain at the price. Not only did he succeed in acquiring a husband for one of his three daughters; it is doubtful if he had paid a single penny by the time he died twenty years later.[54] Yet in the long run it was the ward who was to gain more from the alliance than the guardian. John died in 1336 without male issue. The estates were therefore divided between the three daughters, and Joan and Andrew acquired the manors of Chiddingly, Claverham, Arlington, and Waldern.[55] These all lay conveniently close, in a couple of cases adjacent, to the existing lands of the family and, unlike for example, the manors which came to the Etchinghams by marriage, were to form a permanent addition to the inheritance.

[52] His career is discussed in detail below, pp. 48–51.

[53] *CFR 1307–19*, p. 303.

[54] When he died in 1336, the king ordered that his tenants should not be held responsible for his remaining debts, which included the £120 due for Andrew Sackville's marriage (*CCR 1333–7*, p. 671).

[55] Phillips, p. 85.

But whether that inheritance would continue to descend in the Sackville family was another matter altogether. Joan bore her husband a couple of sons, but both were to predecease their father.[56] A second marriage, to Maud, daughter of John Lovat, brought no further issue, and by the 1360s this ageing knight was beginning to turn his mind to less conventional ways of securing the continuance of his line. For some years past he had found solace in the company of a mistress called Joan Burgess by whom he had sired two children, Thomas and Alice.[57] It was on them that he now decided to settle the estates in default of any surviving legitimate issue. He had enfeoffed his estates in September 1365: thus, by the time he died in July 1369 the feoffees had been in possession for a good five years and their title was secure.[58] In each county where an inquisition *post mortem* was taken the king's escheator heard from the jury of Sir Andrew's dispositions, but could do nothing to challenge them.[59] As for the knight's widow, she was, as far as we can tell, satisfied with the lands settled on her in dower. Or, rather, she and her second husband, Sir Edmund de la Pole, initiated no action in the courts. It fell to the Fawley branch of the family, very belatedly, to mount a challenge when she died in 1393.[60] But it never came to anything. Thomas emerged secured in his title; and his marriage to the daughter of Sir Edward Dallingridge of Bodiam brought him into the orbit of one of the most powerful men in Sussex.

[56] The last mention of Sir Andrew Sackville the younger occurs in June 1365 (*CCR 1364–8*, p. 208). For John, whose approximate date of death is not known, see PRO, JUST 1/941A m. 38d and *CCR 1405–9*, p. 56.

[57] Reviewing S. M. Wright, *The Derbyshire Gentry in the Fifteenth Century*, Colin Richmond asks, 'Would my impression that virtually every Derbyshire gentleman had a mistress and bastards be substantiated?' (*Medieval Prosopography*, 6 (1985), 121). If the Sussex evidence is anything to go by, the answer is yes. For the bastard son of Sir William de Etchingham IV see above, p. 4.

[58] According to the inquisition taken in Sussex the enfeoffment had been made in September 1365 (*CIPM*, xiii, no. 58).

[59] Phillips, pp. 89–90; *VCH Oxon.* viii. 92–3. There was a minor challenge from one John, son of John de Kent, who renewed a long-standing claim of his family's to Claverham and Chiddingly (MS notebooks of Edwin Dunkin, BL Add. MS 39374 fos. 236v–237r).

[60] *VCH Oxon.* viii. 92–3. Oddly enough, what was to dog the family's footsteps right into the fifteenth century was their acquisition of Joan's share of the de la Beche manors. Their title was challenged by the descendants of Isabella, another of the co-heiresses, who married the Oxfordshire knight William FitzElys. The succession to the Sackville estates of an illegitimate son may have encouraged John FitzElys in Henry IV's reign to try his family's luck again; in his own account of the descent of the estates Thomas concealed his illegitimacy by making himself the son of Sir Andrew the younger (G. Wrottesley, *Pedigrees from the Plea Rolls*, p. 260).

III. THE WALEYS FAMILY

Arms: Gules a fess ermine

The story of the Waleys family, by contrast, was to end on a note of failure. In 1436 William Waleys III, the last in the direct line of descent, was declared an idiot and his estates partitioned between his sisters. Ten years later, however, the decision was challenged by William's guardians. The litigation that followed—both between the co-parceners and the guardians and between the co-parceners themselves—dragged on for the best part of the next half-century. Apart from the lawyers, only posterity emerged the richer: for, thanks to the zeal with which the several claimants searched the family archives for evidence to support their arguments, the historian of the medieval Waleys clan is furnished with a wealth of evidence that would otherwise have been lost.[61]

The Waleys lords of Glynde were probably descended from one Godfrey of Malling, who in the 1090s held a hide in the archbishop of Canterbury's peculiar of South Malling which was later to become the manor of Glynde. He may or may not have been the same Godfrey who also held fees at Thannington and Newenden in Kent. The chances are that he was, because a century later all the fees had passed to one Denise, presumably Godfrey's eventual heiress, who married Richard Waleys. Following his death she married again, and Godfrey, her son by Richard, had to sue for the recovery of lands rightly his which the archbishop had granted to his stepfather. He vindicated his title, and the Waleys family were to enjoy possession for the next 200 years.[62]

But not untroubled possession. In the thirteenth century their relations with their archiepiscopal overlords were to be poisoned by the long-running dispute over the terms on which they held the manor of West Tarring, near Worthing. In 1233 Sir Godfrey I held this property at farm for the service of £18 a year or its value in provisions for the archbishop's household when he came to stay. When the

[61] There are good accounts of the family in *The Glynde Place Archives: a Catalogue*, ed. R. F. Dell (Lewes, 1964), ix–xv, and F. R. H. Du Boulay, *The Lordship of Canterbury: An Essay on Medieval Society* (London, 1966), pp. 100–5, 370–3.

[62] The case, which cost Richard Waleys 100 marks, is *Curia Regis Rolls*, vi, 11–12, discussed by Du Boulay, *Lordship of Canterbury*, p. 100. The next two paragraphs are based on this source and Dell, op. cit.

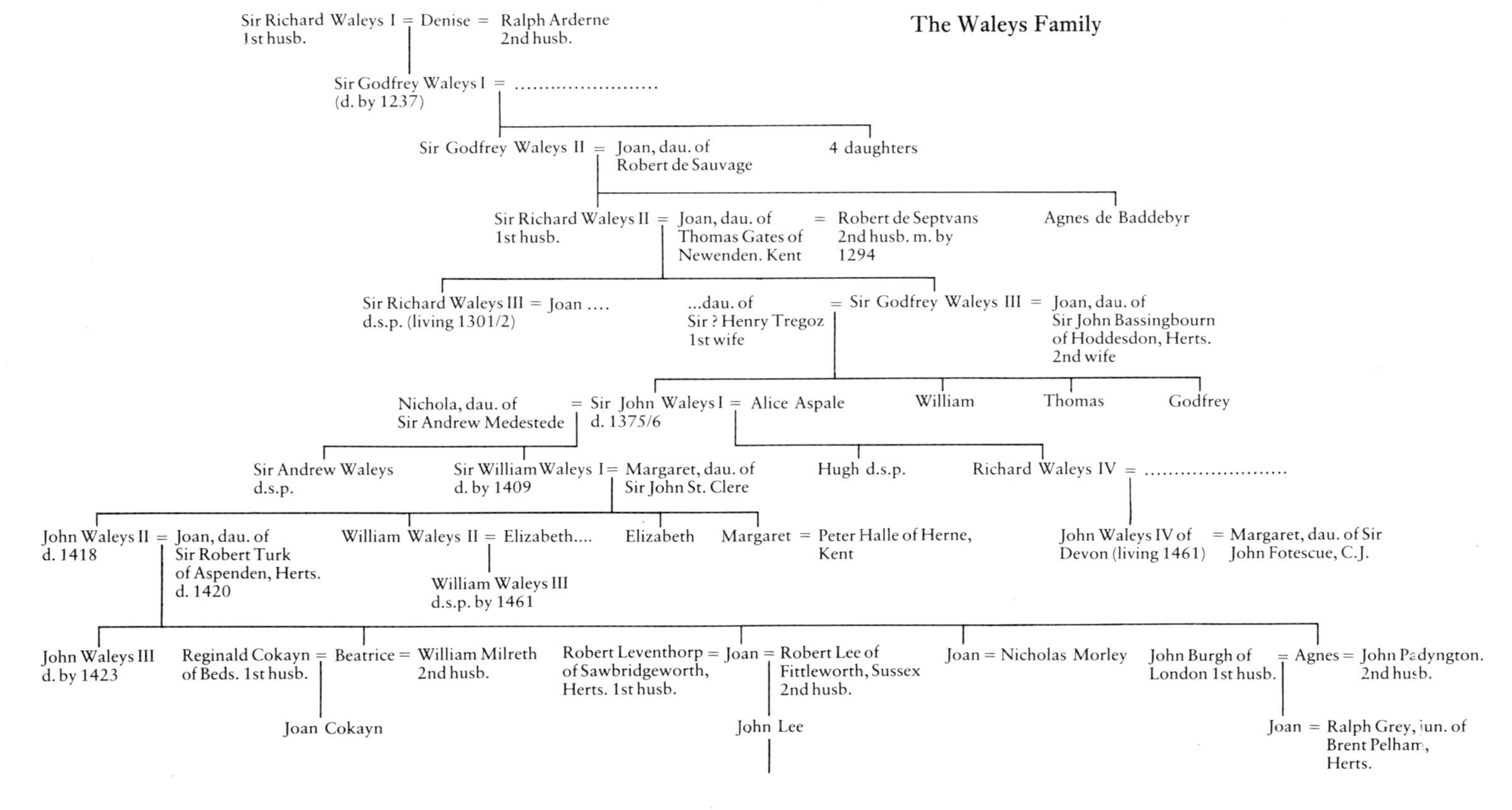

The Waleys Family
Sir Richard Waleys I = Denise = Ralph Arderne
1st husb.
2nd husb.
Sir Godfrey Waleys I = ...
(d. by 1237)
Sir Godfrey Waleys II = Joan, dau. of Robert de Sauvage
4 daughters
Sir Richard Waleys II = Joan, dau. of Thomas Gates of Newenden. Kent = Robert de Septvans
1st husb.
2nd husb. m. by 1294
Agnes de Baddebyr
Sir Richard Waleys III = Joan ....
d.s.p. (living 1301/2)
...dau. of Sir ? Henry Tregoz 1st wife = Sir Godfrey Waleys III = Joan, dau. of Sir John Bassingbourn of Hoddesdon, Herts. 2nd wife
Nichola, dau. of Sir Andrew Medestede = Sir John Waleys I = Alice Aspale
d. 1375/6
William
Thomas
Godfrey
Sir Andrew Waleys
d.s.p.
Sir William Waleys I = Margaret, dau. of Sir John St. Clere
d. by 1409
Hugh d.s.p.
Richard Waleys IV = ...
John Waleys II = Joan, dau. of Sir Robert Turk of Aspenden, Herts. d. 1420
d. 1418
William Waleys II = Elizabeth....
William Waleys III
d.s.p. by 1461
Elizabeth
Margaret = Peter Halle of Herne, Kent
John Waleys IV of Devon (living 1461) = Margaret, dau. of Sir John Fotescue, C.J.
John Waleys III
d. by 1423
Reginald Cokayn of Beds. 1st husb. = Beatrice = William Milreth 2nd husb.
Joan Cokayn
Robert Leventhorp of Sawbridgeworth, Herts. 1st husb. = Joan = Robert Lee of Fittleworth, Sussex 2nd husb.
John Lee
Joan = Nicholas Morley
John Burgh of London 1st husb. = Agnes = John Padyngton. 2nd husb.
Joan = Ralph Grey, jun. of Brent Pelham, Herts.

archbishop was absent Godfrey was bound to treat the tenants without vexation or exaction and as fairly as the archbishop treated his own tenants elsewhere. If he did not, the archbishop could invoke his right to re-enter. In some way Godfrey did offend, and the archbishop, Edmund Rich, took possession. Godfrey must soon have made amends, however, for he was restored in 1237 on payment over four years of £80, which the archbishop agreed to deposit with the prior of Lewes to provide marriage portions for Godfrey's four daughters.

But in 1276 the dispute erupted again. This time the manor was confiscated on the grounds that Richard Waleys had wronged the tenants and had spent only £6. 17*s*. 5*d*. on food when last entertaining the archbishop. Richard countered by claiming that the archbishop had come to Tarring simply as a guest and had not formally been presented with the keys. A day was set for the case to be heard at East Grinstead in July 1276, but it was not one which was going to be resolved easily. Richard's mother challenged the jurors on the grounds that they were all tenants of the archbishop, and the court was adjourned till a fortnight after Michaelmas. The new jury too was challenged, and its successor likewise. But on the fourth attempt, at Westminster in February 1277, an acceptable panel was finally agreed, and judgement given. Richard was found to have exacted services each year from the tenants which he should only have exacted when the archbishop was coming. He was therefore sentenced to forfeit the manor which was thereafter retained in hand by the archbishops.[63]

This Richard was succeeded by his two sons in turn, neither of whom played any part in the political life of their county. Both were summoned to perform military service in the Scottish wars of Edward I and Edward II, but there is no evidence that either performed it.[64] Godfrey III, indeed, was only persuaded to take up the rank of knight after being distrained in 1312.[65] So it was left to his son and heir, Sir John, to rekindle some glory in the family name, and that he did with

[63] Letters patent embodying the judgment were issued (*CPR 1272–81*, pp. 204–7).

[64] The summonses quoted by Dell, p. xi, are summonses and no more. Of the many protections issued around this time in the name of 'Richard Waleys' the only one likely to have been in favour of Richard Waleys of Sussex (and not his Yorkshire namesake) is the one for the followers of John de Warenne, earl of Surrey, in 1298 (Gough, *Scotland in 1298*, p. 44). As there are no horse rolls for Warenne's retinue it cannot be established whether he actually went. For information about military service I am greatly indebted to Mr Andrew Ayton of the University of Hull.

[65] PRO, E370/1/13 m. 7.

gusto.[66] Knighted before 1340 he took part in all the main campaigns of his time. He went on the Crecy–Calais expedition of 1346–7 in the retinue of Sir Michael de Poynings, accompanied Roger Mortimer, earl of March on his *chevauchée* in Picardy ten years later, and finally in 1359–60 joined Edward III on the great winter campaign which was to bring this phase of the Hundred Years War to its conclusion.[67] In the years of peace that followed, he settled down at home to shoulder the burdens of office-holding. He was sheriff in 1364 and a knight of the shire in parliaments from 1368 to 1371. By then he was probably a very old man. According to the guardians of William Waleys III in the following century, he was 'an C (hundred) yer olde' when he died, in January 1376, 'as yt ys welknoen to gret partie of the schyre of Sussex'.[68] This reckoning was based, as Dell observed, on the curious premiss that, as he was bedridden for more than a year before he died, he must have been 'gretly passede the said age of lx and xi (71), for he was a sclendre mane and also lusty a knyght in ys dayes as was any of ys age wythine the reyme of Ingelond'.[69] Admittedly it suited the guardians' convenience to be able to claim that Sir John had enjoyed a life-span well beyond that normally given to a medieval knight. But the tradition to which they appealed—that he lived, if not to a hundred, then at least to a ripe old age—was in fact one which was more firmly rooted than their eccentric line of reasoning might lead us to suppose. Sir John had performed homage for his Kentish manors back in 1315 and 1318; on the assumption that he did so, as was usual, on attaining his majority, he could surely have been little short of eighty by the time he lay bedridden in the mid-1370s.

Three days before he died he surrendered his estates to a group of feoffees—John Whitecliff, vicar of Mayfield, Robert de Ore, John Cockfield, and John Preston of Glynde.[70] He was not, like many who did likewise, worried about the responsibility of a wardship. His

[66] Perhaps that is to put it too politely. He was a rather lawless character as well. See below, pp. 73–4.

[67] He was knighted by 18 June 1340 (GLY 1139, no. 74). For the Crécy campaign see *Crécy and Calais*, ed. G. Wrottesley (London, 1898), p. 85, and for Picardy, PRO, C76/33 m. 10. Sir Andrew Sackville also fought under Mortimer's banner in 1355: he was a retainer of his. Could it therefore have been he, acting as a sub-contractor, who recruited Waleys? For the campaign of 1359–60 see PRO, C76/38 mm. 5, 113.

[68] GLY 24.

[69] Ibid. For the long dispute between William Waleys III and his cousins, in connection with which this deposition was made, see below, pp. 89–90.

[70] GLY 19 (i).

problem was a different one. Andrew, his eldest son, had gone on a pilgrimage to Jerusalem and had not yet returned.[71] His father wanted to protect his interests; and this he did by requiring the feoffees to hold the estates in trust for the duration of his absence and to settle them on him when he returned. By 1378, however, there was still no sign of him, and it was doubted if he could still be alive. Yet the feoffees could not act because if he was dead they could not make the enfeoffment, and without the enfeoffment the remainders to Andrew's brothers could not take effect. To overcome this legal conundrum they decided to make another settlement, this time on two clerks, Giles Wringlesworth and John Sadler, for a period of two years, at the end of which they would grant them to Andrew with remainders successively to his brothers William, Hugh, and Richard.[72] The remainders were then presumably invoked and William allowed to succeed.[73]

The new lord of Glynde was a man of similar mould to his father; at least he was closer in spirit to his father than, say, to his shy, retiring grandfather. It is doubtful if he ever saw active service in France;[74] but then the opportunities for waxing rich in his day were not what they had been a generation earlier. In his native county, however, he was a busy office-holder: he served twice as sheriff, in 1383 and 1395, and was five times elected an MP. He also served from time to time on various commissions.[75] But by the late-1380s the management of his estates was coming to loom ever larger in his preoccupations.

The difficulty that landowners experienced in making demesne husbandry pay in this period has become a commonplace among historians. The combination of high costs and low profits was enough to persuade a good number of them to abandon direct cultivation altogether. William Waleys and his neighbours, however, were not going to give up without an effort.[76] They retained their demesnes in hand until the 1380s and later, but in William's case, when the day of

[71] GLY 19 (iii).

[72] GLY 4 and 19 (iii).

[73] GLY 1139, fo. 1, no. 2, is a grant by the feoffees to Sir William Waleys of the Kent manors only.

[74] This is not to rule out the possibility that he fought before succeeding to the estates, in which case he would probably not have bothered to obtain letters of protection. The William Waleys whose name appears among the esquires in the retinue of Sir Reginald Cobham in 1387 (PRO, E101/40/33) could not have been this man, for the latter had been knighted at least five years earlier (GLY 4a). Perhaps it was his second son, William II—or a member of a different family altogether.

[75] *CFR 1381–5*, pp. 249, 591; *CPR 1377–83*, pp. 147, 187.

[76] For a discussion of the survival of demesne husbandry see Ch. IV.

reckoning came, it was not just demesnes but whole manors that were then placed under indirect management. In July 1386 he demised the manor of Patching to the local vicar and four other men for a period of six years at an annual farm of £40. The arrangement broke down two years later when he accused them of failing to pay the instalment of rent due at Easter 1389.[77] He won his case, and in October 1392 found three new farmers to take on the estate. In a recognizance made before the mayor of Chichester they acknowledged a debt to the knight of £150. 16*s*. 10*d*. which they agreed to pay out of the issues of Patching over three years at the rate of £36. 6*s*.8*d*. per annum. Shortly before the lease was due to expire, however, Sir William was back in the courts—in King's Bench this time—alleging that the lessees had defaulted in payment of the final instalment.[78] The outcome of the case is not known, because the parties came to a settlement out of court. But whatever the terms of that settlement may have been, the question remains: what was Sir William trying to achieve by these leases? If his objective had been to eliminate the losses incurred by direct husbandry, he could have done so quite simply by leasing the demesne alone to a single lessee. The presence of several lessees and the registration of the second lease at Chichester suggest rather that Waleys may have been demising Patching as security for a loan. But in that case, surely, he would have acknowledged *his* indebtedness to *them*. And why did the repayments provided for come to only £109 when the recognizance was for £150. 16*s*. 10*d*.?

The arrangements that Sir William made for some of his other manors were more conventional. In 1394 he leased Hawksden, less its park, for nine years to one Geoffrey Aleyn for an annual rent of £3. 14*s*. 4*d*.[79] And in 1403 he leased Thannington in Kent for four years to John Bossington and John Pecham, both of Canterbury, for an annual rent of £20.[80] These demises were a straightforward response to the collapse in the profitability of demesne husbandry that took place in the last quarter of the fourteenth century. Sir William decided to cut his losses by exchanging the hazards of direct management for the security, or relative security, of a rental income.[81] These early leases were only granted for short periods—nine years and four years

[77] PRO, CP40/523 m. 123d.
[78] PRO, C44/21/17; KB27/545 m. 37; *CCR 1396–99*, pp. 56–7.
[79] GLY 1223.
[80] GLY 1341.
[81] See below, Ch. IV.

respectively. Sir William was allowing for the possibility that, should prices rise, he might want to re-enter. But once it became clear that they were not, it may be presumed that he would have granted renewals for much longer periods.

If some manors were demised for reasons of economic and administrative convenience, others were used to endow junior members of the family. For example, there was Sir William's younger half-brother Richard to be taken care of. He had been granted, probably by his father, an annuity of ten marks to be received out of the issues of Glynde, Patching, and Bainden. But like anyone else in his position he aspired to an estate of his own. So in September 1382 his elder brother granted him the manor of Bainden for his (the donor's) lifetime in part payment of the annuity.[82] At the same time there was Sir William's eldest son and heir John, who had been growing restless since his marriage in the mid-1390s to Joan, daughter of Sir Robert Turk. In June 1398 his father settled on him and his heirs the manors of Glynde and Patching, and three years later those of Hawksden and Bainden too.[83] William must have been subjected to quite exceptional pressure to part with so many properties, particularly when Glynde itself was the principal seat and residence of his family. But part with them he did, and not apparently without regret. The row simmered on until 1406 when it was resolved by the appointment of arbitrators. The old man recovered Glynde and Patching; the son was assigned Hawksden and Bainden.[84] Bainden was of course the very manor that Sir William had given to Richard in 1382 in part payment of his annuity, and on which John too had had a claim since 1401. Richard may well have remained in possession, however, while the wrangling dragged on, because John leased the estate to him again after succeeding his father in 1409.[85]

Whether or not there was much brotherly love in the Waleys family, there was certainly no shortage of enmity between father and son, uncle and nephew. And the reason was quite simply competition for land. All three of the families we have met were well-to-do ones by the standards of the medieval gentry. The Waleyses were not quite so well

[82] PRO, CP40/573 m. 352d. I have relied on Edwin Dunkin's notes from this roll in BL Add. MS 39375, as the original is now deemed unfit for production.

[83] GLY 5, 6; PRO, CP40/562 mm. 106–106d. The latter must be referring to the reversion of Bainden.

[84] GLY 1140 (v). The episode is discussed again in Ch. III in the context of disputes which arose within, as opposed to between, landowning families.

[85] GLY 1235.

endowed as the Etchinghams and Sackvilles, but even they could muster some five or six manors. In 1347/8, on the eve of the Black Death, Glynde itself was worth some £80 gross or £45 net. Liveries of cash recorded on the account roll for that year give some idea of the likely value of the other properties.[86] £20 were paid over by the reeve of Patching, £9 by the reeve of Thannington and £11. 13*s*. 0*d*. by the reeve of 'Southall'.[87] These figures are obviously incomplete—that for Thannington, for example, refers to the issues of two terms only. Moreover, there were also the profits of Bainden and Hawksden to be taken into account. The rental income of the former came to £6. 17*s*. 0*d*. in 1363.[88] Perhaps we ought to think in terms of a level of net receipts of £120 per annum or more on the eve of the Black Death and, if the rents paid by the farmers of Bainden and Thannington are any guide, a little less than that half a century later.

A figure of this order represented a not insubstantial income for a knightly family. Many would have had to subsist on a good deal less.[89] But whether or not it was adequate to meet all the demands likely to have been placed upon it is another question altogether. For, as Michael Bennett has written, a gentry estate 'was subject to continous fragmentation and reunification through temporary grants and leases to widows, younger sons, trustees, and others'.[90] The history of the Waleys family at the turn of the fourteenth and fifteenth centuries illustrates clearly one such period of fragmentation. A widow, an ambitious young heir, and a representative of the half-blood were each in possession of a part of the inheritance. Sir John's widow held the dower third to which she was entitled by common law. John II, Sir William's eldest son, was granted a couple of manors by his father on (or shortly after) the occasion of his marriage in the late-1390s. And Richard, a younger son of Sir John by his second marriage, was granted an estate in Bainden by

[86] GLY 1072.

[87] Southall has disappeared now. It may have been in Selmeston ('Proofs of Age of Sussex Families', *SAC* 15 (1863), 211). There are accounts for the manor from 1417 to 1421 in PRO, SC6/1029/19.

[88] GLY 1117.

[89] A quarter of all the gentry families resident in Gloucestershire had, on the evidence of the Nomina Villarum of 1316, only a single manor (N. E. Saul, *Knights and Esquires: The Gloucestershire Gentry in the Fourteenth Century* (Oxford, 1981), p. 225). Lionel de Bradenham, an Essex gentleman whose estate Dr Britnell has discussed, held just one manor, Langenhoe ('Production for the Market on a Small Fourteenth-Century Estate', *Ec.H.R.* 2nd ser. xix (1966), 380–7).

[90] M. J. Bennett, *Community, Class and Careerism: Cheshire and Lancashire Society in the Age of Sir Gawain and the Green Knight* (Cambridge, 1983), p. 87.

his elder brother in lieu of a cash annuity. This last piece of generosity was to cost Sir William dear in later years when it led to litigation between himself and his heir. But, combined with the other temporary alienations, it was also to cost him dear in the short run in the sense of receipts lost to the main household. The Waleys estates, although not insubstantial by comparison with those of many other gentry families, were not wide enough to support two, let alone three, establishments for any length of time.[91] No wonder, then, that Sir William proved so anxious to recover Glynde and Patching from his son.

In truth the head of any gentry family found himself compelled to reconcile two conflicting demands. On the one hand, he had to make the best provision he could for children other than the legal heir; on the other, he had to preserve the integrity of the inheritance and to transmit it intact from one generation to the next. For most gentry families it was the latter instinct that triumphed, and not surprisingly. Given their limited resources they had no alternative. We know that in one county—Gloucestershire—nearly a third of all the resident gentry families held two manors or fewer.[92] These people and their like in other counties would have found it simply impossible to make adequate provision for their younger offspring. The implications for English social history are worth pondering. England never developed a French-style 'noblesse'. So the younger sons of the nobility, broadly defined,[93] could only cling on to their status so long as they had means to support it. Some found ways of doing so, as we shall see;[94] but others did not. For their sakes let us not overlook the effects of downward as well as upward mobility as a solvent on the divisions in English society.

Our three Sussex families were, however—to repeat—all well endowed. The half-dozen or dozen manors they held apiece enabled them to do more for their younger sons than most proprietors of their rank were able to do. In that respect they can hardly lay claim to being regarded as representative. Nevertheless an examination of the settlements they made of their estates may shed some light on the extent to which those knightly families which had the means felt obliged to provide for younger kin and on the ways in which such provision could be made.

[91] The estates were certainly supporting three establishments between 1398, after John's marriage, and 1409, when he succeeded his father.

[92] See above, p. 21 n.

[93] Taking the nobility to embrace both the parliamentary peerage and those whom we now call the gentry.

[94] See below, Ch. II, *passim*.

A landowner's freedom to do what he would with his properties was, of course, severely limited by the rules governing their descent. These required all land held in fee simple by the tenant on the day that he was alive and dead to pass to his son and heir. Only chattels, not lands, could be devised by will. But in the late middle ages landowners found a way of circumventing this restriction: they granted their lands to a group of feoffees to be held in their 'use' for the grantor's lifetime and to be disposed of after his death according to instructions given in his last will. Here was a neat means of both devising lands by will and excluding the lord from wardship in the event of a minority. The nobility, as we know from the work of Mr McFarlane and Professor Bean, used it to the full.[95] Younger sons were looked after better than ever before or afterwards; and cadet branches of great houses flourished. But by 1500 the position of the eldest son and heir had once again been made secure. For this he had to thank his future father-in-law, who had no desire to see a daughter enter into a depleted inheritance.[96]

By the fourteenth century most if not all the magnates employed a combination of the use and the entail to control the descent of their estates. And so too very likely did the gentry. Indeed, it has been suggested that during the period of their emergence uses were employed both at an earlier date and to a greater extent among mesne tenants—that is, mainly the gentry—than among the tenants-in-chief.[97] There can be no doubt that at the end of that period they were employed as commonly by the knights as by the higher nobility, both to evade feudal incidents and to provide for younger sons. A total of as many as 140 are recorded in inquisitions *post mortem* taken between 1327 and 1377.[98] This figure represents, of course, only those made by tenants-in-chief, whose lands were seized by the king's escheator. It does not take into account the many more that were made by mesne tenants. For the most part these are now lost to view, but their existence is confirmed by the evident concern that tenants-in-chief felt at the erosion of their revenues. Sometimes they took measures to arrest it. On 24 February 1352, for example, Queen Philippa secured the appointment of a commission of oyer and terminer to investigate trespasses committed on her estates in southern England, and it is evident from the surviving

[95] McFarlane, *Nobility of Later Medieval England*, pp. 60–82, 268–78; J. M. W. Bean, *The Decline of English Feudalism, 1215–1540* (Manchester, 1968), chs. 3 and 4.

[96] McFarlane, *Nobility of Later Medieval England*, p. 278.

[97] Bean, p. 134.

[98] I have totted up the lists in Bean's Appendix II.

roll of indictments from Sussex that she must have instructed her justices, doubtless all hand-picked, to be on the watch especially for collusive enfeoffments.[99] In the rape of Pevensey, for example, the queen laid claim to the wardship of Philip Medsted, a minor, whose father William had enfeoffed Sir John Waleys and Ralph Pulscote eight days before he died. She also claimed the wardship of young William Etchingham, whose father James, shortly before embarking for France, had enfeoffed his brother Master John and John de Ore.[100] Fear of death abroad and the sheer cost of equipping a contingent for active service should be numbered among the reasons that led a proprietor to grant his estates to feoffees. But they are not the only reasons, or even the main ones. If James de Etchingham could have set out secure in the knowledge that he had a son of mature age to succeed him, the precaution of enfeoffement would scarcely have been necessary. In other words, the primary impulse in the making of enfeoffments—and this Queen Philippa seems to have realized—was less the concern to provide for younger sons than to exclude the chief lord from exercising his or her right to wardship.

This may be a bold conclusion to draw, given that our knowledge of the instructions that Medsted and Etchingham gave to their respective feoffees derives entirely from the record of the oyer and terminer commission. But it is one which is consistent with the evidence we can glean from other sources. This suggests that the knights had no need to

[99] A commission was issued initially on 18 Feb. 1352 to John de Molyns, John de Lewknor, William de Notton, Richard de Cressevill, John de Roches, and John Claymond, 'touching the persons who broke Queen Philippa's parks and closes at Bristol, Redcliffe, Kingswood ... and elsewhere, hunted in the parks and warrens, felled her trees, fished in her several fisheries ...' (*CPR 1350–4*, p. 288). The emphasis is clearly on illegal hunting and fishing. But in a second commission, issued a week later on 24 Feb., the terms of reference were drawn far more widely, and given to a different group of justices—John de Molyns, John de Lewknor, William de Notton, Peter de Richmond, Thomas de Ingleby, John Claymond, John Knyvet, and Edmund de Chelrey were told not only to investigate territorial trespasses but also 'withdrawal of rents, escheats, wards, marriages and other profits pertaining to her ...' (ibid., 287). The great seal was moved by a writ under the 'secret seal'. Sir John de Molyns was Queen Philippa's steward of the household (N. M. Fryde, 'A medieval robber baron: Sir John Molyns of Stoke Poges, Buckinghamshire', in *Medieval Legal Records*, edited in memory of C. A. F. Meekings by R. F. Hunnisett and J. B. Post (London, HMSO, 1978), p. 206).

[100] PRO, JUST 1/941A mm. 29d, 32. The activities of the queen's justices form an interesting comment on Edward III's kingship. If Edward, as we are told, was willing to comply with the wishes of his tenants-in-chief by the granting of licences for enfeoffment to use, the same could hardly be said of his wife in respect of her own tenants. In her keenness to maximize her income from land she reminds one of an earlier queen, Edward I's Eleanor of Castile.

assign feoffees to provide for their kindred because they had already done so themselves. The fourth Sir William de Etchingham, for example, had seen that none of his three brothers went without some means of support. Robert, who already held two manors *iure uxoris*, was granted a life interest in a third, at Padbury (Bucks.).[101] Simon, the next brother, was tonsured and provided with the rectory of Herstmonceux, and Richard, the youngest, was given lands in the Romney Marsh villages of Snave, Brookland, and Ivychurch in tail male, with remainder to the grantor and his heirs.[102] When we recall that Sir William had already granted a couple of manors to his friend Sir Alan Buxhill, he may be said to have shown himself generous—generous almost to a fault.[103] In a sense he could afford to be, as he had no legitimate sons to provide for.[104] But if either or both of the two junior lines had taken root, the losses to the inheritance could have been long-lasting if not permament. Fortune, however—or, rather, genetics—came to the rescue. Robert proved no more productive of issue than his elder brother had been, and in the late 1330s the patrimony was reconstituted on the succession of Richard's son James. Ironically, the factor of infertility which had encouraged dispersal in one generation promoted reunification in the next.

Generous though he undoubtedly was towards his brothers, Sir William had taken care to minimize the loss of income to himself. He preserved intact the main core of manors around Etchingham itself, and endowed his kin with the smaller manors further afield.[105] Lords generally parted with the latter more willingly than they did with the former. In the early-1300s Etchingham's contemporary Sir Richard Waleys III of Glynde established his brother and eventual heir Godfrey at the manor of Thannington, fifty miles away near Canterbury.[106] Later in the fourteenth century, as we have seen, Sir William Waleys was making provision for *his* younger brother, Richard. Initially he had been given, as one suspects many younger sons were, a cash annuity, in

[101] For Padbury see *VCH Bucks.* iv, 211. Robert de Etchingham and his wife settled the manors of Glottenham (Sussex) and Holwist (Kent) on the heirs of their bodied with remainder to the right heirs of Petronilla (*Feet of Fines for the County of Sussex, 34 Henry III—35 Edward I*, ed. L. F. Salzman (SRS vii, 1908), no. 1127).

[102] 'Kent Fines, Edward II', ed. J. Greenstreet, *Arch. Cant.* xiii (1880), 295.

[103] The apparent 'grants' to Buxhill need not, of course, have been 'grants'. They could have been sales; or they could have been exchanges. It is possible as well that they could have been a form of retaining fee.

[104] However, he is known to have had a bastard, for whom see below, p. 000.

[105] This is consistent with his decision not to retain the manors brought to him by his wife.

[106] GLY 24.

this case of ten marks; but in 1382 this was partially converted into an interest for the donor's lifetime in the Wealden manor of Bainden.[107] Richard, like most of his peers, preferred land to cash, for land brought status as well as security. But if he thought he was being fobbed off with land that was of less than equivalent value, he would have had little hesitation in coming back to ask for more. This was just what Sir Thomas Sackville's son and daughter-in-law did in 1408 after he had settled on them the twin manors of Newland and Southey. They persuaded him to grant them additionally an annuity of £10 out of the issues of Chalvington; but he did so only on the condition that they and their heirs *quietly* held the manors he had already granted them.[108] He had given way, one senses, with a little ill grace.

Evidently, then, there was a concern for the well-being of the family as a whole—provided that the family is defined narrowly enough to exclude collaterals. Younger sons were endowed as far as means would allow, but less than might have been expected through the medium of the use. Despite the advantage it might appear to have offered of endowing younger sons at the expense of their elder brethren rather than at the expense of their father, the use was employed less to divide the inheritance than to ensure its survival in the event of an awkward succession. We have seen how uncertainty over the whereabouts of his eldest son and heir led the aged Sir John Waleys to grant his estates to feoffees to hold until such time as Andrew returned.[109] Here Sir John was at least seeking merely to protect and secure the rights of the heir at common law. A few years earlier, however, Sir Andrew Sackville employed feoffees to convey his estates to a bastard son who had no rights at all at common law. The reason, as we have seen, was that he had no legitimate surviving sons of his own. Yet, so strongly did he feel the burden of dynastic responsibility, that he did all in his powers to ensure the survival of the family name through the succession of his bastard. This tactic was by no means unique. The Durham chronicler tells us that William, the last of the de Vesci lords of Alnwick, who died in 1297, had conveyed his estates to Anthony Bek, bishop of Durham, 'trusting in him that he would keep them to the use of his little illegitimate son

[107] See above, p. 000. For a discussion, useful for comparative purposes, of provision for younger sons in fifteenth-century Derbyshire see S. M. Wright, *The Derbyshire Gentry in the Fifteenth Century* (Derbyshire Record Soc. viii, 1983), pp. 30–1.

[108] BL Add. MS 39490, fol. 61.

[109] GLY 19 (iii).

William and surrender them to him when of age'.[110] William de Vesci's trust was misplaced. His feoffee let him down. Sackville, however, was more successful. From Thomas, his son by Joan Burgess, were to descend the Sackvilles of Knole, lords Buckhurst and earls, later dukes, of Dorset, one of the proudest dynasties of Stuart and Georgian England.

[110] Bean, *Decline of English Feudalism*, pp. 118–19.

II

THE LOCAL COMMUNITY

THE meeting that took place at Etchingham's house that day in May 1366 can stand for many others that took place at all levels of landed society in medieval England. The attestation of a deed was a service that gentlemen regularly performed for one another. It was an expression of solidarity, perhaps of friendship, and a contribution towards the cohesion of the local community.[1] It provides a convenient point of departure, therefore, for a discussion not only of the families themselves but also of their friends and neighbours. Some of them are better known than they themselves are—Sir Edward Dallingridge, the builder of Bodiam, for example, or Sir John Pelham, Henry IV's councillor and the founder of a great Sussex dynasty, who makes an appearance at the very end of the century. These men and others we will meet in so far as they impinged on the lives of the Etchinghams, Sackvilles, and Waleyses. But prosopography only rises above the anecdotal or antiquarian if it is wedded to an analysis of social institutions; for these dictate the lines along which relationships and connections are formed and acted out. Let us begin, therefore, by looking at the structure of lordship in fourteenth-century Sussex.

The lands of our three families lay in the rapes of Hastings and Pevensey, the two easternmost of the six such territorial lordships into which Sussex was divided. Both rapes were held in the fourteenth century by non-resident proprietors. The former was granted in 1268 to John, son of the duke of Brittany, a son-in-law of Henry III, whose descendants held it for the next three-quarters of a century. After the family's extinction in the direct line it was resumed by Edward III and settled in 1342 on his three-year-old son, John of Gaunt, who held it with the earldom of Richmond until 1372. In that year diplomatic considerations led Edward III to grant it again to the duke of Brittany, John de Montfort, but after the latter's change of sides nine years later it was seized and granted by Richard II to his queen, Anne.[2]

[1] M. J. Bennett, *Community, Class and Careerism*, p. 32.
[2] VCH Sussex, ix. 2–3.

The rape of Pevensey, immediately to the west, was usually held by the consort of the reigning monarch—by Margaret in the early fourteenth century and by Philippa in the middle decades. After the latter's death in 1369, it was settled by the widower king on his son John of Gaunt, now duke of Lancaster, in exchange for the surrender of Hastings. Though again an absentee, Gaunt was by no means a man to let his interests in the county go by default. He had to endure considerable provocation from a group of local landowners led by Sir Edward Dallingridge who were determined to prevent his officials from holding a hundredal court at Hungry Hatch, near Fletching. For as long as Dallingridge enjoyed the support of his patron, the earl of Arundel, there was little Gaunt could do, but in June 1384, when the earl was in temporary eclipse at court, he struck back. Dallingridge was attached to answer charges brought both by bill and by indictment. He was found guilty on all counts and fined. He even suffered brief imprisonment. But time was to show that he had gained more than he had lost by the use of violence. Gaunt may have won the right to hold a court at Hungry Hatch; but in practice he chose to abandon it. The goodwill of a powerful knight mattered more than the suit of court owed by a few hundred tenants.[3]

These periodic assertions of lordship—Queen Philippa, as we have seen, had sent her own justices of oyer and terminer into the rape of Pevensey a generation before—rarely had any lasting effect.[4] Absentee lordship, one suspects, was little different in practice from no lordship at all. Immediately west of the Ouse, however, it was a different story. Here the Warenne earls of Surrey had long held sway. The rape and castle of Lewes formed part of the original endowment the first William de Warenne had received from William the Conqueror. Though stretching across no fewer than thirteen counties, it fell into

[3] S. Walker, 'Lancaster v. Dallingridge: a Franchisal Dispute in Fourteenth-Century Sussex', *SAC* 121 (1983), 87–94. For a more detailed discussion of this revealing episode see below, p. 75.

[4] For a discussion of the indictments heard by Sir John de Molyns and his fellow justices in 1352 see above pp. 23–4. How much contact was there between a foreign overlord and his tenant? If the answer is in general very little, they did have to meet sometimes to go through certain formalities. In 1309, for example, the Robertsbridge Chronicle tells us that the abbot journeyed to Fotheringhay Castle (Northants.), where John, duke of Brittany was residing, to persuade him 'non sine lacrimis' to give his licence to the proposed alination to the house by Sir William de Etchingham of the advowsons of Salehurst, Udimore, and Mountfield (BL Add. MS 28, 550, fol. 1^r–1^v, summarized in C. S. Perceval, 'Remarks on some Charters . . .', *Archaeologia*, xlv (1979), 430).

four main blocks of land, centred on Lewes in Sussex, Reigate in Surrey, Castle Acre in Norfolk, and Conisbrough in Yorkshire.[5] The earl and his household moved around their estates, as all the nobility did. But if any of their properties could lay claim to precedence in their affections it was probably Lewes: there, on the crest of the ridge, lay one of their strongest castles, and at its foot the great priory church of St Pancras where they were laid to rest. It is possible that in the early fourteenth century, as the focus of government moved northwards in the wake of the Scottish wars, they came to spend more time on their Yorkshire estates; but it is by no means certain.[6] And the splendid barbican that John, the last earl (d. 1347), built in front of the south gate at Lewes is a measure of the importance that he still attached to the satisfactory upkeep of his Sussex property.

Lordship of the rape of Lewes gave the Warennes a powerful instrument of control over their locality. But it was no substitute for winning the support and co-operation of the local gentry. This they sought to do both formally and informally—formally by taking on some of them as retainers, informally by making common cause as far as possible with the 'class' as a whole. The informal workings of lordship are by their very nature impossible to analyse. The absence from east Sussex in the early-fourteenth century of any major feuds between the earl and rival landowners may or may not be a tribute to their effectiveness—it depends on the confidence one has in the value of negative evidence.[7] The formal ties of retaining, however, are a different matter. Although no actual indentures of the Warenne family have come down to us, membership of the retinue can be reconstructed by making an analysis of letters of protection and witness

[5] GEC xii. i, 493, gives the Conqueror's grant. The family's Sussex manors formed a compact block within their rape of Lewes: Cuckfield, Clayton, Ditchling, Newhaven, Brighton, Patcham, Rottingdean, Houndean, Northease, Rodmell, Keymer, Middleton, Allington, Pyecombe, Seaford, Ilford, Piddinghoe, and, further north, Worth.

[6] And not just the focus of royal government: the last earl's feud with Lancaster over the abduction of the latter's wife led to the harrying of his Yorkshire estates (J. R. Maddicott, *Thomas of Lancaster, 1307–22* (Oxford, 1970), pp. 207–8). On the other hand, in the 1330s he seems to have spent more time in the south than in the north. Of the four charters of his from that decade noted by F. R. Fairbank, 'The last earl of Warenne and Surrey', *Yorks. Arch. Jnl*. xix (1907), 193–264, three were issued at Lewes and only one at Hatfield (Yorks.).

[7] This statement is based on a perusal of the rolls of the Court of King's Bench. In the first half of the fourteenth century the only such feud to reach the court was that between Sir John Waleys and a group of the earl's retainers acting on his authorization (see below, pp. 35 and 73–4).

lists to charters. The former reflect the temporary increase in numbers occasioned by the demands of military service; the latter are probably a better guide to the peacetime composition of the affinity. But the point is not important: whichever source is consulted, we experience no difficulty in identifying a dozen or more dependants whose constant attendance on the earl in the early fourteenth century attests their position as permanent retainers. Roughly half of them were Sussex men: Sir William Paynel, Sir Simon Pierpont, Sir Luke de Vienne, Sir Nicholas Gentil, Sir Robert Stangrave, and Sir Thomas Poynings and his brother Michael. The remainder—the Neirfords, Thomas and John, Sir Ralph and Sir John Bigod, Sir Constantine Mortimer, Sir John de Wisham, Sir Nicholas de Malemains, and Sir Richard de Hakelut—came from elsewhere.

The Neirfords, a Norfolk family, hailed from the manor of Narford, a few miles west of Castle Acre. The proxmity of their estates to those of the Warennes made them natural allies of the earls.[8] But their prominence in the retinue in the first half of the fourteenth century owed as much to the liaison between earl John and their kinswoman Maud as it did to considerations of political geography. Where the two first met is not known. It could have been in Norfolk. But it could equally have been at court where Maud's husband, Sir Simon de Dryby, was a household knight of Edward II.[9] However and wherever the romance began, it certainly blossomed, and by 1316 the earl was initiating divorce proceedings against his wife, Joan of Bar. But Joan was of royal blood, and an annulment was out of the question. By the late 1320s her husband was probably resigned to taking her back again.[10] If Maud fell from favour as a consequence (and the evidence is, to say the least, ambiguous), her kinsmen certainly did not. Thomas and John were important figures in their own right in Norfolk society, and the former was witnessing Warenne charters as late as the 1340s.[11]

Two other knights could lay claim to Norfolk connections. In the

[8] John and Thomas de Neirford appear in protections (PRO, C81/1741/63 and 67; *CPR 1321–4*, p. 237) and as charter witnesses (*CPR 1313–17*, p. 653). Sir Thomas died in 1344 holding manors in Houghton, Holt, Panworth, and Narford (Norfk.). He and his wife also held the manor of Sedlescombe (Sussex) from the Prior of the Order of St John of Jerusalem for their lives (*CIPM* viii, no. 508). This is likely to have been a consequence not a cause of the connection with the Warennes.

[9] It is Fairbank, 'The last Earl of Warenne and Surrey', 198. who identifies Sir Simon as her husband. He appears as a household knight of Edward II in the following wardrobe books: BL Add. MS 17362 fol. 55^v; Add. MS 9951 fol. 25^r, Stowe 553 fol. 102^r.

[10] Fairbank, 230.

[11] *CPR 1340–3*, p. 512.

reign of Edward II there was Sir Constantine Mortimer, a tenant of the Warenne family from whom he held the manor of Attleborough. His father had died while he was still under age, and his upbringing as a ward of the earl explains why he began his adult career in the retinue.[12] A little later there were the two Bigods, father and son, and bearers like Mortimer of a distinguished noble name. They were members of that junior line of the Bigod earls of Norfolk which was dispossessed when the childless earl Roger surrendered his title and estates to Edward I and received them back entailed on the heirs of his body.[13] Stockton, near the Suffolk border, was their principal seat.

Sir John de Wisham came from a rather different background. He was probably a west Midlander by birth.[14] His parentage is obscure, but by the time of his death in the 1330s he had become a rich man and a figure of some importance in the affairs of the realm. A good marriage had helped to get him started;[15] membership of the royal household took him further. He was an esquire of the king by 1303 and a knight by 1313.[16] He was primarily what in the twelfth century would have been called a 'curialis'.[17] Throughout the time he was serving John de Warenne he was on King Edward II's payroll; and the further connections he forged with the nobility were with magnates prominent at court. In the 1320s, for example, he is found accompanying the king's half-brothers Edmund, earl of Kent, and Thomas, earl of

[12] GEC ix. 243, 248–9. He accompanied the earl to Boulogne in January 1308 for Edward II's marriage (ibid.), and was present at the Dunstable tournament in 1309 (A. Tomkinson, 'Retinues at the Tournament of Dunstable', *EHR* lxxiv (1959), 75). From 1314 he is found in the service of Aymer de Valence, earl of Pembroke (J. R. S. Phillips, *Aymer de Valence, Earl of Pembroke, 1307–24* (Oxford, 1972), pp. 257–9, 302), and after his death that of the younger Despenser (E. B. Fryde, 'The Deposits of Hugh Despenser the Younger with Italian Bankers', *Ec.HR*, 2nd Ser., iii (1951), 362).

[13] For Earl Roger's settlement see McFarlane, *Nobility of Later Medieval England*, p. 262. The manor of Settrington (Yorks.) was their consolation prize, as he puts it. Sir Ralph Bigod is found serving the earl in the late 1320s and 1330s (PRO, Calendar of Ancient Deeds A.313, A.10886; CP40/416 m. 321; *Lewes Cartulary*, i. 69). For Sir John Bigod, who was serving in the 1330s and 1340s, see PRO KB27/352 m. 26 and 354 m. 109d.

[14] His main seat was probably Woodmanton, in Clifton-on-Teme (Worcs.), for which he was granted a licence to crenellate in 1332 (*CPR 1330–4*, p. 242). The timber-framed chapel which still survives may date from this time (N. Pevsner, *Worcestershire* (Harmondsworth, 1968), p. 122).

[15] On 9 Oct. 1303 he was granted permission to marry Eleanor, widow of William de Ferrars (*Calendar of Documents Relating to Scotland*, ed. J. Bain (Edinburgh, 1887), ii. 359).

[16] *Cal. Docs. Scot.* ii, p. 359; PRO, E101/375/8 fol. 35^r.

[17] As described, for example, by E. Mason, 'Magnates, Curiales and the Wheel of Fortune', in *Proceedings of the Battle Conference II, 1979*, ed. R. A. Brown (Woodbridge, 1980), pp. 118–40.

Norfolk; and his career reached its height when he served under the former as seneschal of Gascony during the Saint-Sardos War.[18] Sir Nicholas de Malemains was also a royal household knight at roughly the same time that he was in Warenne's employment.[19] He was a member of a family that held lands in Kent, but his own manors, except one in Surrey, all seem to have lain in Lincolnshire.[20] Sir Richard de Hakelut was a kinsman of Sir Edmund and Sir Walter de Hakelut, two more of Edward II's household knights;[21] but he himself is not recorded as a member of that affinity. Like Wisham he seems to have hailed from the west Midlands—at least, from the counties which border the Welsh Marches. That apart, not much is known about him.

The remaining retainers were all Sussex men. The most important were the two Poynings brothers, who were members of a local family of baronial rank and who may very likely have led a sub-retinue in the way that, for example, the Berkeleys did in the affinity of the earl of Pembroke.[22] Sir William Paynel was also an important and active man in his own right, and owned a string of manors across southern England.[23] Sir Simon Pierpont, a landowner in both Sussex and Suffolk, was a tenant of the Warennes in the two manors which he held at Newick and Hurstpierpoint, respectively north and north-west of Lewes.[24] Sir Luke de Vienne was the younger brothers of Sir Peter de Vienne, lord of Bilsham and Cudlowe in the west of the county near Arundel.[25] Sir Nicholas Gentil was by all accounts another knight of

[18] See for example *CPR 1321–4*, pp. 399, 403, and *The War of Saint-Sardos (1323–1325)*, ed. P. Chaplais (Camden 3rd Ser. lxxxvii, 1954), p. 271. Other details of Wisham's career are heaped together in Moor, *Knights*, v. 203–4.

[19] He was admitted on 26 May 1306 (PRO, E101/369/11 fol. 106). He is found in Warenne's service at roughly the same time (Tomkinson, 'Retinues . . .', 75; *CPR 1307–13*, p. 406; PRO, C81/1741/63).

[20] He held Rothwell, Langton and Toynton (Lincs.). His Surrey manor was Ockley (Moor, *Knights*, iii. 98).

[21] Sir Walter Hakelut was admitted as a knight of the royal household on 25 May 1306, the day before Malemains (PRO E101/369/11 fol. 106). Sir Edmund had followed in his footsteps by 1313 (E101/375/8 fol. 35^{r}). For what little is known about Sir Richard see Moor, *Knights*, ii. 171.

[22] For a more detailed discussion of the Poynings family see below, pp. 37–8.

[23] He is found in the earl's service between 1308 and his death in 1317 (Moor, *Knights*, iv. 21–2). He was one of the earl's nominees on a commission of oyer and terminer to investigate trespasses on his estates in 1312 (*CPR 1307–13*, p. 531).

[24] He is found in the earl's service from before 1317 (*Lewes Cartulary*, i. 55) until 1342 or later (*CPR 1354–8*, p. 18, being an enrolment for a grant, dated 1342, from the earl to 'his bachelor' Simon Pierpont). For his manors see Moor, *Knights*, iv. 51 and *VCH Sussex*, vii. 175.

[25] One Sir Luke de Vienne, probably the father of the retainer, paid £3. 12*s*. 3*d*. in taxation at Bilsham and Cudlowe in 1296 (*Three Earliest Subsidies*, p. 79). For the careers

local origins. He was three times sheriff of Surrey and Sussex and twice as often its knight of the shire in parliament; his arms, moreover, are under Surrey and Sussex in the Parliamentary Roll of Arms.[26] But, oddly enough, there is no record of where his lands lay. Nor are we any better informed about the land-holdings of Sir Robert Stangrave. He served as sheriff of the same two counties for the twelve months 1310–11, but his appointment obviously owed far more to the influence of the earl of Surrey, who nominated him, than it did to any independent standing he had in his own right.[27] His coat of arms does not appear in any surviving roll of arms; and his manors, if any, cannot be identified.

Doubtless there were other local men connected with the earl who have escaped our grasp. William de Northo, several times a sheriff and knight of the shire, is one possibility. Sir William de Braose, lord of Bramber, is another. Both men witnessed charters occasionally.[28] But when all due allowance is made for possible omissions, these still do not look quite the body of dependants an earl of the Warenne family might be expected to have taken into his pay. Some, like Thomas and Michael de Poynings, were men who would have earned a place in any retinue; but others, like Gentil, Stangrave, and Hakelut, had a good deal less to recommend them than such well-established and apparently better-qualified local knights as William and Robert de Etchingham and Richard and John Waleys—to name only the most obvious examples—with whom Earl John seems on the evidence at our disposal to have had little or no connection at all. Sir William de

of both Lukes, senior and junior, see Moor, *Knights*, v. 123–4, and for the latter's connection with Warenne, PRO, C81/1741/65 and 67.

[26] He was a member of the retinue by 1309 (Tomkinson, 'Retinues . . .', 75; *CPR 1307–13*, pp. 405, 406; PRO, C67/16 m. 7, E101/17/31). By 1330 he was also favoured by Edmund, earl of Kent, who wrote in that year to the Chancellor asking for Gentil and Ralph Camoys to be appointed keepers of the peace in Sussex (A. Harding, *The Law Courts of Medieval England* (London, 1973), p. 97). For a summary of his career see Moor, *Knights*, ii. 108. He was a knight, and evidently had no difficulty in supporting the rank. But where could his lands have been? The only pointer is the sum of 2*s*. 4*d*. he paid in the parliamentary subsidy of 1327 at West Hampnett (*Three Earliest Subsidies*, p. 126), but Sir John St John was the lord of the manor there. Perhaps he was a household knight, in the strict sense of that phrase.

[27] For his appointment at the request of the earl see PRO E159/84 m. 27d. He witnessed a charter of the earl in 1309 (*CPR 1307–13*, p. 406) and accompanied him to the Dunstable tournament (Tomkinson, 'Retinues . . .', 75). He is not mentioned by Moor. For obscure retainers serving as sheriffs and the means by which they obtained office see Saul, *Knights and Esquires*, pp. 155–7.

[28] For these two as charter witnesses see *CPR 1307–13*, p. 406, and *Lewes Cartulary*, i. 55. For Paynel's career see Moor, *Knights*, iv. 21–2.

Etchingham IV was at least known to him, perhaps even on friendly terms with him: he had witnessed some charters of his, and had been made welcome at Reigate castle.[29] Sir John Waleys, on the other hand, was an outright enemy. In 1338 he had been arrested, so he said, on the earl's authority by his illegitimate son, Sir William de Warenne, and then imprisoned for eight weeks in Lewes castle. Fourteen years later he was still to be pursuing his captors in the courts.[30] All the same, when he heard of the earl's death in 1347, he laid aside these animosities and, with all the other local notability, attended the funeral at St Pancras's Priory; indeed, he invited the abbot of Battle and Sir Thomas de Pashley to stay overnight with him at Glynde before setting out on the final stage of the journey to Lewes.[31] Some of the mourners, John included, had but recently returned from the siege of Calais. The importance which they and others not directly of the earl's affinity attached to being present that day gives some insight into the sense of loss felt by a community that had known his lordship for as long as forty years. His influence in Sussex society was not to be measured exclusively in the number of life-retainers who took his livery.

In the second half of the century Lewes was no longer the centre of lordship it had once been. When John de Warenne died, his estates passed to his nephew Richard FitzAlan, earl of Arundel, who resided chiefly at Arundel in the west of the county and at Shrawardine in the Welsh Marches.[33] Lewes was a useful acquisition; but it was peripheral to the family's main concerns. In 1377 it was left undefended against the French, and four or five years later it could not even withstand an assault by a gang of local malefactors protesting against the continued exaction of labour services.[34] Its days as a residence had passed; it was used principally as a gaol. But one wonders if it was ever secure enough to keep prisoners in.

Richard FitzAlan was already a rich man before his uncle died, and

[29] *CPR 1313–17*, p. 653.

[30] See below, pp. 000.

[31] GLY 1072, 'Avena'.

[32] *Crecy and Calais*, ed. G. Wrottesley (London, 1898), pp. 85, 128, 137, for the presence of Sir John Waleys and other Sussex men in the retinue of Sir Michael de Poynings on this expedition.

[33] For the FitzAlans see R. R. Davies, *Lordship and Society in the March of Wales, 1282–1400* (Oxford, 1978), pp. 56–8. John de Warenne's widow continued to hold the estates until her own death in 1361, after which they passed to Arundel. However, he was already renting the lands in the Welsh Marches from her (*CPR 1345–8*, pp. 434–7).

[34] A. Goodman, *The Loyal Conspiracy: the Lords Appellant under Richard II* (London, 1971), pp. 110–11.

the acquisition of his estates made him richer still.[35] He and his son could afford to spend generously on retaining fees; and they did. Their affinity was probably larger than the Warennes' had been; it was certainly represented in more English counties. Cambridgeshire provided Sir Payn Tiptoft, Cheshire Sir Hugh Browe, Wiltshire Sir John Dauntsey, Northumberland Sir William Heron, and Northamptonshire Sir John Fallesle.[36] But as far as southern England was concerned, it was Sussex that was the heartland of the family's interest. There they could rely on Sir Edward St John of Stopham, Sir Thomas Camoys of Trotton, Sir William Percy of Petworth, Sir Thomas Poynings of Dedisham, and the brothers Sir Roger and Sir Edward Dallingridge, all of them men who were as prominent in the service of the FitzAlans as they were active in the affairs of their county.[37] In the distribution of their landholdings they reflected the firm location of the comital family's power-base in the west of the county.[38] But neither the elder nor the younger Richard FitzAlan allowed his preference for living at Arundel to stand in the way of cultivating the goodwill of the gentry of east Sussex. Sir Edward Dallingridge brought with him into the family's fold his friend Sir Philip Medsted and his son-in-law Sir Thomas Sackville.[39] Sackville, in fact (though he may not have realized it), was following in the footsteps of his father, who had served the elder Richard FitzAlan in the 1340s; but there is no reason to doubt that his own recruitment was the consequence of his kinship with Edward Dallingridge.[40]

Once again we have to remember that the surviving sources give only an incomplete picture of the composition of the retinue. There are other knights whom we may suspect, but not prove, to have been within its fold: for example, Sir William Waleys, who received a

[35] A good idea of the wealth of the elder Richard FitzAlan is given by the inventory of his goods and stocks of cash drawn up after his death in 1376 (L. F. Salzman, 'The Property of the Earl of Arundel, 1397', *SAC*, 91 (1953), quoting BL Harl. MS 4840 fos. 393–5). Goodman's estimate that the younger Richard FitzAlan's income in 1397 totalled £3,700 should be regarded as the lowest possible figure.

[36] Goodman, pp. 114–21.

[37] Ibid., and the references there cited. Sir William Percy was an MP for Sussex no fewer than twelve times. Goodman notes Sir Edward Dallingridge but not his elder brother Sir Roger who was a feoffee of the elder Richard FitzAlan (*CPR 1364–7*, pp. 237–8). For both men see below, pp. 67–8.

[38] The only two exceptions are the Dallingridges. Roger lived at Sheffield Park (in Fletching) and Edward, of course, at Bodiam.

[39] Walker, 'Lancaster v. Dallingridge', p. 89. On 25 July 1384, at Arundel's request, Philip Medsted obtained from the king a pardon for a felony (PRO, KB27/502 Rex m. 11).

[40] See above, p. 13.

pardon for supporting the Appellants in 1387–8.[41] But already enough evidence has been produced to show that the FitzAlans had at their disposal a powerful range of connections in Sussex and beyond. In the eastern half of the county, certainly, aristocratic influence waned neither as far nor as fast as the decay of Lewes castle might lead us to suppose.

The Warennes and their successors the FitzAlans were the only families of comital rank to reside in Sussex in the fourteenth century. To that extent they would have been regarded as the natural leaders of gentry society in the county: but not necessarily the only leaders. There were other local families whose heads were issued with personal summonses to parliament, but not usually over more than a generation or two.[42] The Etchinghams, as we have seen, enjoyed the honour for just a single generation.[43] The Tregoz family of Dedisham enjoyed it for two.[44] The Poynings clan did rather better: they enjoyed it for no fewer than four. The distinction was not undeserved. They had the wealth to support their rank; and the martial prowess of successive heads of the family brought lustre to their name.[45] They were indeed, as Round once said, a warrior race.[46] They took part in all the main campaigns of the fourteenth century; but few of them lived to enjoy the fruits of victory. Sir Michael died at Bannockburn in 1314, his eldest son at Honnecourt near Cambrai in 1339, and his second son at Calais in 1347.[47] In terms of survival their followers did better

[41] Goodman, pp. 38, 43.

[42] I am taking a personal summons to parliament (i.e. to sit with the Lords) as the criterion of baronial rank.

[43] See above, p. 3.

[44] GEC xii. ii. 24–8; L. F. Salzman, 'Tregoz', *SAC* 93 (1955), 34–57. The witnesses on the roll of Tregoz charters (BL Add. Roll 56358) give an idea of their range of contacts. Sir Michael de Poynings crops up, as he does on Warenne charters. Another regular witness is Sir William de Wolvercote, who was an MP for Sussex in 1338, 1339, and 1346.

[45] GEC x. 659–64, gives an account of the family. The manors held by Sir Michael de Poynings at the time of his death in 1369 were Poynings, Pangdean, Perching, Hangleton, West Dean, Twineham, Waldron, Crawley, and Slaugham (Sussex); Terlingham, Newington Bertram, Westwood, Eastwell, and Horsmonden (Kent); Wrentham (Suffolk); Wilton (Norfolk); and La Lee and La Gore, by Lavington (Wilts.) (*CIPM* xii, no. 404). This Sir Michael had been the first member of his family to receive a summons to parliament. But he was the heir to a great reputation: he had been allowed to inherit his lands while still a minor in consideration of the merits of Sir Thomas de Poynings, his father, who met his death in the attack on Honnecourt (*CPR 1338–40*, p. 395).

[46] J. H. Round, 'The Lords Poynings and St John', *SAC* 62 (1921), 1.

[47] GEC x. 659–64. The list is not exhaustive: Sir Richard was to die on Gaunt's expedition to Spain on 25 May 1387.

than they themselves did. Sir John Waleys, who accompanied Sir Michael and his uncle to Crecy and Calais, lived, so his descendants claimed, to be 100, and Sir Roger Dallingridge, a companion-in-arms on the same expedition, lived to be a septuagenarian.[48] Certainly, the fame of Sir Thomas and his brother and son and the sheer frequency with which the went off to fight had the effect of attracting to their service a small group of knights who accompanied them on campaign after campaign. Whether or not these war-time contingents ever constituted a peace-time retinue in the accepted sense is hard to say; but, in so far as they did, they seem to have formed a sub-retinue within the wider Warenne affinity. The Poynings lords, as we have seen, were frequently in the company of the last earl, and they and the knights who fought with them are likely to have supported rather than challenged his authority and that of his FitzAlan successors.[49]

If we want to look for providers of good lordship who lay outside the Warenne–FitzAlan network of clientage, we would do better to look to the ecclesiastical magnates than to the minor secular barons. In eastern Sussex the archbishop of Canterbury probably counted for a good deal more than the latter did. The nature of his relationship to the local gentry is, however, difficult to establish. He could assemble an affinity, but not lead it in war. He could demand the performance of knight service, but not don armour himself. He could marry off a ward, but not wed her himself. His ability to lead and to forge alliances with powerful neighbours may to that extent have been limited; but the influence he exerted over the lives of his tenants was nevertheless real enough. His estate was the largest in south-eastern England, and his position as a territorial magnate was buttressed by the franchisal powers he wielded in his domains.

One of the oldest endowments of the see was the great manor of South Malling, which stretched from Lewes north-eastwards to the county boundary twenty miles away. Like other such estates in Kent and Sussex it was divided into territorial tithings called 'boroughs', and these in their turn fell into two groups: Wadhurst, Mayfield, Greenhurst, Framfield, and Uckfield being designated 'within the

[48] For Waleys see GLY 24. Roger Dallingridge was said to have been aged 24 or more when his father died in 1335 (*CIPM* vii, no. 649). He was still alive in January 1377, when he served as an MP for the last time.

[49] Among the knights who fought with them regularly were Sir Andrew Peverel (PRO, C81/1736/102; *Crecy and Calais*, ed. Wrottesley, p. 138) and Sir Roger Dallingridge (PRO, E36/203 fol. 126v; C76/15 m. 21; C76/17 m. 19; *Crecy and Calais*, p. 137). Once again I am grateful to Andrew Ayton for references.

wood', and Wellingham, Gote and Middleham, Norlington, Ashton, Southerham, South Malling, and Stonham 'without the wood'.[50] At the time of Domesday Book the whole of this vast manor was farmed for the sum of £90 by one Godfrey of Malling, who is likely to have been the ancestor of the Waleys family of Glynde. The connection is not entirely beyond doubt; but it is a fair assumption given the facts as we have them. By the end of the twelfth century the hide in South Malling (probably Glynde) which Godfrey had held as a tenant of the archbishop had passed to a certain Denise, whose first marriage was to a Richard Waleys. The difficulty which Godfrey, their son, had in establishing his right to succeed illustrates well the power which the archbishops had over the lives of their tenants. Denise married again, and the archbishop assigned to her second husband lands which Godfrey claimed should have passed to him. He won his case, but only at great cost to himself.[51]

Glynde, as we have seen, was the principal residence of the Waleys family. But it was not their only possession: they also held Bainden and Tarring (Sussex) and Thannington (Kent), with Glynde, from the archbishop for the service of 2$\frac{3}{4}$ fees.[52] Thus they must have been one of the very few families to hold all of their manors, spread over two counties, from the archbishop and no other lord. Usually the buying and selling of land and the consequent lengthening of the tenurial chain had the effect of weakening the bonds of tenure at the same time as the growing self-awareness of the gentry made their links with each other seem all the more important.[53] But in the case of the Waleys family there seems little reason to doubt the continuing importance of the ties with the archbishop. When Sir Godfrey III succeeded his brother in 1303, he went to Aldington in Kent to do homage for his fees and to pay the relief of £15.[54] His son Sir John was present at Charing in 1350 when Archbishop Islip granted the keeping of Slindon Park to his huntsman John le Venour.[55] And it was entirely appropriate that when his successor Langham resigned the see in 1368 Sir John should

[50] F. R. H. Du Boulay, *The Lordship of Canterbury: An Essay on Medieval Society* (London, 1966), pp. 124–5.

[51] See above, p. 14, and *Curia Regis Rolls*, vi. 11–12. It cost him 100 marks.

[52] This was in 1210–12 (Du Boulay, pp. 370–3, 377). Tarring, of course, was later lost.

[53] On this theme see also below, pp. 57 and 153.

[54] *Register of John Pecham, Archbishop of Canterbury, 1279–1292*, ed. F. N. Davis and others unnamed (Cant. and York Soc. 64, 1969), p. 19.

[55] *CPR 1358–61*, p. 195.

have been appointed one of the keepers of the temporalities during the vacancy.[56]

The extent to which the archbishop's influence in east Sussex extended beyond the ranks of his own tenants is less easy to ascertain. He visited the county often enough, certainly, because he had country residences—perhaps it would be better to call them palaces—at Slindon and Mayfield.[57] But whom did he see, whom did he invite to dinner, when he was staying there? The names of some of his guests are recorded on a couple of household rolls of the 1340s preserved among the muniments of Westminster Abbey. They are mostly great men of the realm: the earls of Huntingdon, Arundel, and Northampton, Sir John Pulteney, Sir Bartholomew Burghersh, and the bishop of Chichester. A few knights appear—Sir Roger Cobham, Sir Arnold Durford—but with the possible exception of Sir Geoffrey le Say there are no Sussex men.[58] Of course, we do not know whether these people were being dined at Mayfield, Lambeth, or somewhere else: the likelihood is that it was at Lambeth. If we knew with whom the archbishop dined or—better still—with whom he went hunting at Mayfield, a very different picture might emerge. For whatever the regard in which *he* held the men of Sussex, they certainly held him in very high esteem indeed. In 1297, when they were summoned by the sheriff to elect two knights of the shire, they refused to do so on the grounds that the archbishop of Canterbury and others, bishops, earls, barons, and knights, were absent from the county court.[59] In acknowledging the leading role the archbishop played in the formation of opinion, the county electors were merely recognizing the reality of the power which he wielded locally. Twenty years earlier, when issue had been joined between the archbishop and his wayward tenant Sir Richard Waleys, two successive juries had had to be discharged because Richard and his mother alleged that they were composed of the archbishop's tenants.[60] A proprietor with as large a stake in a society as the archbishop had in

[56] *CPR 1356–68*, pp. 327, 390.

[57] Mayfield is described by I. Nairn and N. Pevsner, *Sussex* (Harmondsworth, 1965), p. 566, as having 'one of the most spectacular medieval halls of England'. An early-fourteenth century date is suggested. It is illustrated on plate 32a. Slindon, near Arundel, was where Stephen Langton died (F. M. Powicke, *Stephen Langton* (London, reprinted 1965), p. 160).

[58] Westminster Abbey Muniments 9222, 9223.

[59] J. R. Maddicott, 'The County Community and the Making of Public Opinion in Fourteenth-Century England', *TRHS* 5th Ser., 28 (1978), 31.

[60] *CPR 1272–81*, p. 206: 'because all the jury were of the liberty of the archbishopric, Joan refused to proceed'. For a discussion of this case see above, p. 16.

Sussex could scarcely have avoided exerting an effortless influence on the outcome of pleas to which he was a party. The manipulation of local politics more generally, however, would have required the intervention of his stewards—the men, that is, who would have represented him at sessions of the county court. Centrally he employed two: one having responsibility for the lands, the other responsibility for the liberty. The holders of the latter office would have been the men with legal experience. Few of them, it seems, hailed from Sussex, and not many were scions of gentry families.[61] But one exception was Sir Edmund de Pashley, who is found serving as steward of the liberty in 1301/2, quite early in his career.[62] That said, however, it seems likely that the archbishop's central stewards would have been less visible in Sussex than the stewards of his great manor of South Malling. These would have been men with considerable influence locally; but in the absence of any surviving court rolls from there only one of them can be identified by name—a certain John Preston, who lived at the very end of the century.[63] His predecessors in the office must sadly remain the *eminences grises* of our story.

The archbishop was the most important ecclesiastical proprietor in eastern Sussex. But three others need to be taken into account: the bishop of Chichester, the abbot of Battle and the prior of Lewes. Of these the last two were for our purposes the more important.

The prior of St Pancras, Lewes enjoyed the distinction of being head of the premier Cluniac house in England. The community had been founded in the 1070s by William de Warenne and his wife Gundrada. Staying at Cluny while on their way to Rome they had been greatly impressed by the quality of life of the monks and sought permission to establish a colony on their lands in Sussex. However, Abbot Hugh, who did not like 'transmarine ventures', was unenthusiastic, and it was only after further entreaties from William that his reservations were overcome and the community finally established.[64] One concession which the founder had wrung from Hugh was that Cluny would always appoint as prior of St Pancras the holiest

[61] I base this judgement on the list of office-holders in the archbishop's temporal administration in Du Boulay, Appendix A.

[62] *Registrum Roberti de Winchelsey Cantuariensis Archiepiscopi A.D. 1294–1313*, i, ed. R. Graham (Cant. and York Soc. 51, 1952), p. 414. Three years earlier he had been a bailiff of the liberty of the archbishop (PRO, E401/143). I owe this reference to Dr Paul Brand.

[63] *CPR 1396–9*, p. 205.

[64] D. Knowles, *The Monastic Order in England, 940–1216* (Cambridge, 2nd edn., 1963), p. 151. The phrase quoted is at p. 154.

and wisest monk available, excepting the grand prior of Cluny itself and the prior of La Charité-sur-Loire.[65] Lanzo, the first prior, more than lived up to expectations, and subsequent appointments too were of high quality.[66] But, all the while, it was implicitly understood that whomever Cluny selected should meet with the approval of the patronal family. As Archbishop Pecham wrote to the abbot in 1285, since the earl is patron, 'you are bound by the laws of gratitude to comply with him as far as you can without offence to God; and do not believe that the monastery can prosper in spiritualities or temporalities if he is offended'. However, on this occasion the archbishop's advice went unheeded and the earl *was* offended. He wanted an Englishman; and in the event he was given a Frenchman—and an unworty one at that.[67]

The choice of a prior was a matter of concern to the earl precisely because relations between his family and the house were so close. They had been its principal benefactors; they remained its surest friends; and when they were absent abroad they usually made its prior one of their attorneys.[68] The last earl had had his disagreements with the monks; but he too showed that he bore them no lasting ill-will by asking, as his ancestors had, to be buried within their walls.[69] The surprise would have been if he had chosen otherwise. What is perhaps more remarkable is that so many members of gentry families who were by no means near neighbours should have chosen to be buried there as well: for example, Sir John Arundel (d. 1379), Katherine, Lady Peverell, the widow of Sir Thomas Poynings (d. 1409), and Sir Edward St John and his wife.[70] Clearly the monks had lost little of the popularity they had enjoyed in their early years. What cost them dear in the fourteenth was not the financial difficulties that were apt to afflict late-medieval monastic communities but the complications arising from the war with France.[71] As an alien priory their house was

[65] Ibid., 151. [66] Ibid., 153.

[67] S. Wood, *English Monasteries and their Patrons in the Thirteenth Century* (Oxford, 1955), pp. 64, 166.

[68] For Joan of Bar, the last earl's widow, appointing the prior as an attorney see *CPR 1354–8*, pp. 34, 222, and *1361–4*, p. 5.

[69] There had been trouble, for example, in 1314, when the earl's bailiff at Reigate was ordered to desist from harming the priory's property (*VCH Sussex*, ii. 67). The only Warenne earl not buried at Lewes was William III, who died in 1148 while returning from the Second Crusade (GEC xii. i, 497).

[70] *Testamenta Vetusta*, ed. N. H. Nicolas (2 vols., London, 1826), i. 97, 105, 120; GEC x. 662, and xii. ii, 27.

[71] *VCH Sussex*, ii. 66–9, summarizes the problems.

liable to have its possession seized whenever hostilities broke out. Denization, which was granted in 1351, overcame that problem; but it could not deliver the community from the constant danger of raids by the French. The prior, like other local landowners, was compelled to assume a measure of responsibility for the defence of the coastline. But the role suited him ill: in 1377 he tried to repulse a force of French that had landed at Rottingdean, but having only inadequate forces at his command was himself taken captive.[72]

No such fate was ever allowed to overcome the redoubtable Hamo de Offington, abbot of St Martin's, Battle, from 1364 to 1383. He too found himself in the front line in this period; but he assumed the mantle of leadership with greater enthusiasm than did the prior of Lewes. Enjoined 'to array and equip with arms all the men in the county capable of defending it, so as to resist hostile invasion and the destruction of the English tongue', he assembled a force to defend Winchelsea, and when challenged by the French to pay a ransom for the town retaliated by offering battle and putting them to flight.[73] Hamo was equally incisive in his attitude to the internal regime of his house. In the 1360s he instituted radical economy measures intended to reduce its substantial indebtedness. He abandoned direct management in favour of leasing most of the demesne manors; and he took the revolutionary step of cutting the abbey's legal expenses from between £48 and £59 annually to a mere £13.[74] As Dr Searle has written, this amounted to abandoning the liberty which his predecessors had taken such pains to build up.[75]

In its fully developed form in the early fourteenth century the liberty of Battle was the product of a couple of centuries' evolution. But its origins lay in William the Conqueror's initial endowment. For the immediate security of his new foundation he had granted it a banlieu or 'leuga' of all the lands lying within a league of the high altar. Within this roughly circular area, a mile and a half in radius, the abbot was sole lord and judge.[76] The intention had probably been to free him from the jurisdiction of the counts of Eu, who were themselves the sole authority in the rape in which Battle was sited.[77] It was not so much a

[72] *Chronicon Monasterii Sancti Albani Thomae Walsingham, Historia Anglicana*, i, ed. H. T. Riley (Rolls Ser. 28, 1863), 342.
[73] Ibid. 341, and E. Searle, *Lordship and Community: Battle Abbey and its Banlieu, 1066–1538* (Toronto, 1974), p. 342. The royal commission is *CPR 1377–81*, p. 360.
[74] Searle, pp. 263–4.
[75] Ibid. 263.
[76] Ibid. 23, 197.
[77] This and the next paragraph are based on Searle, pp. 197–245.

delegation of royal authority as a protection against the local feudatory. But in the course of time it developed into something more. It could not remain static because the world around it was changing. In the thirteenth century reforms in the office of sheriff, the creation of new officials like the coroner and the intrusion of the eyre justices all had the effect of compelling the abbot to re-define, to clarify, perhaps even to expand, the privileges he had inherited from his predecessors. In 1253 he obtained from the king a charter enabling him to appoint his own coroner and to collect all fines and amercements imposed within the liberty by a royal justice: identification and collection of the profits of criminal justice due to him was thereby considerably facilitated. But he still had difficulty in making good his claim to the profits of civil pleas. These he was as fully entitled to collect as the profits of criminal business, but he found difficulty in so doing because, unlike the very greatest franchise-holders, he did not enjoy the right to appoint his own justices whenever the eyre justices came into the county. A solution was found in 1271 when Henry III conceded that one of the monks could sit on the bench alongside the eyre justices to watch over his community's interests. This was not only a guarantee that the monks could collect the profits due to them but also an enhancement of the dignity of their house: in effect, they were allowed to keep separate rolls for the liberty.

By the early fourteenth century the task of defending the abbey's jurisdictional privileges had called into being at Battle a class of monks well versed in legal precedent. They combed their own and the king's archives for charters of land purchases and proofs of exemption from the Statute of Mortmain which they brought together in a new cartulary, the first for nearly a hundred years. Several of the monks who would have used it are known to us by name. The two most important were John of Whatlington, steward in the 1290s and later abbot, and Henry of Rye, steward from 1307 until about 1327. John of Whatlington, according to Dr Searle, was an administrator of roughly the generation and—it might be added—roughly the stamp of Canterbury's Henry of Eastry and Westminster's Walter de Wenlok: able and autocratic, and a champion of high-farming. But the age of the monk stewards was in fact to be short-lived: a mere forty years. Before and after Brothers John and Henry the holders of the office were laymen.[78] For all their skills the monks of Battle could no more

[78] Ibid. 246.

dispense with the services of the local gentry than could any other house with interests to protect.

The stewards employed by the abbey after 1330 have been characterized by Dr Searle as 'young men of well-connected gentry families of Sussex or Kent, at the commencement of professional careers in law and administration'.[79] Careerism, if not professionalism, may indeed be said to have been common to them all. Yet the ways in which their careers developed varied considerably. Most of them would have had a legal education of sorts. But only one, Sir Robert Belknap, pursued a career in the central courts.[80] The others were content to remain in local practice. Robert de Sharnden, the first of them, came from a family of sub-knightly rank with holdings near Mayfield, and for at least a part of the twenty years he was in the abbot's service was also a bailiff of the liberty of the archbishop of Canterbury.[81] William de Pagham, who followed him, evidently came from west Sussex; but he was only a stop-gap appointment. In 1352 he was succeeded by Robert Belknap, who was to serve for the next quarter of a century. It is doubtful if this ambitious man could have found the time to attend closely to his responsibilities at Battle once his responsibilities in the courts of common law accumulated, and in the later years of his term a very different man is sometimes found serving in his place. This was Roger Ashburnham, younger brother of John Ashburnham, of the village of that name, and the builder of Scotney Castle. He was a soldier and administrator, and someone to whom we shall return.[82] But Battle was peripheral to his interests, and before long he had been replaced by one John Brook. He was a man in Sharnden's mould[83]—a very minor gentleman whose qualifications and innate ability commended him to a number of employers. As well as steward of the abbot of Battle he was at various times steward of Sir John de la Warr, advocate of the bishop of Chichester and perhaps,

[79] Ibid. 419–20.

[80] His career is summarized by Searle, pp. 420–1.

[81] Ibid. 420; but the reference to him as the bailiff of the liberty of the archbishop is *CCR 1333–7*, p. 44, and not as there cited by Searle. Sharnden Old Manor is about a mile and a half east of Mayfield, and its dependent lands abutted on to the park which Sir John Waleys constructed at Hawksden, immediately to the south, in 1337 (AMS 5896/5 m. 2).

[82] See below, pp. 68–9.

[83] Searle, p. 422, describes him as 'a man of the pattern of Sharnden and Belknap'. I hesitate to compare him to the latter. Belknap was an altogether more powerful man—after all, he became Chief Justice of Common Pleas. Brook and Sharnden may well have served as attorneys at Westminster; but they were never professional pleaders there.

too, the adviser of other lords not unknown to us.[84] Robert Oxenbridge who succeeded him was also a man on the make. From the abbey's point of view he was a less satisfactory appointment than Brook, because he was so often away that the beadle had to take over the representation of the house in the royal courts and at the Exchequer. The reason was simply that, like his predecessor, Oxenbridge served several employers simultaneously; and in his own estimation the one who took priority was not the abbot of Battle but that rising star at court, Sir John Pelham of Laughton.[85]

In perceiving how his interests could best be served he was probably right. He could do without the monks of Battle; and they could do without him. Before long they had found the man who was to serve them for longer than anyone else—Bartholomew Bolney. Appointed in the 1420s he was to hold the office of steward until almost the end of his days in 1477. His fortunes were closely bound up with those of the abbey, but not so closely as to preclude service with other local employers. The lands he bought all marched with those of Battle;[86] but the connections he forged with great lords like Archbishop Stafford served to make him more widely known in county society as a whole. Not surprisingly, then, he was appointed to commissions of the peace over a period of no less than twenty-five years. In fact he is another example of what is by now emerging as a recognizable type: the successful estate official who builds up a modest fortune and mixes private with public service.[87] Bolney's career was anticipated by Oxenbridge's and his in turn by Brook's. Of the three Oxenbridge may be judged the most successful in

[84] Searle, pp. 421–2. In 1396 Brook was a feoffee of Sir John Dallingridge (*CCR 1392–6*, p. 499) and twelve years later a feoffee of Sir Thomas Sackville (BL Add. MS 39490 fol. 61). He was an MP for Sussex in 1393. His career is described by A. Rogers, 'The Parliamentary Representation of Surrey and Sussex, 1377–1422', Nottingham Univ., MA thesis, 1957, pp. 190–4.

[85] Searle dismisses Oxenbridge briskly on p. 422. There is more to be found on him and his family in W. D. Cooper, 'Notices of Winchelsea in and after the Fifteenth Century', *SAC* viii (1856), 214–15, and M. Clough, 'The Estates of the Pelham Family', Cambridge Univ., Ph.D. thesis, 1956, pp. 177 ff. This man was probably the builder of the first Brede Place (Rape of Hastings Architectural Survey, Report no. 381, a copy of which is lodged at the National Monuments Record, London).

[86] Searle, p. 422. For Bolney's lands see *The Book of Bartholomew Bolney*, ed. M. Clough (SRS, 63, 1964).

[87] The type so illuminatingly discussed by R. A. Griffiths, 'Public and Private Bureaucracies in England and Wales in the Fifteenth Century', *TRHS* 5th Ser. 30 (1980), 109–30.

fourteenth-century terms because he founded a dynasty.[88] Bolney, like him, built up an estate, but had only a daughter to whom he could transmit it.[89]

If dynastic success was denied him, however, he could look back with satisfaction on a record of public service which compared more than creditably with those of any of his predecessors. In the course of an exceptionally long life he had served on more commissions than they; but his career had developed along roughly the same lines. It began with nomination to minor offices and commissions; and culminated in appointment to the commission of the peace.[90] Bolney, as we have seen, was on the bench for no less than a quarter of a century. Brook and Oxenbridge made it there in the last years of Richard II's reign.[91] Interestingly, they were not the only estate officials to be appointed. John Preston, a fellow justice on the same commission, was the archbishop of Canterbury's steward of South Malling and Otford.[92] And Thomas Blast was an employee of the FitzAlans.[93] These four were the only 'mere' gentlemen (if we might so describe them) named in the two Sussex commissions for 1397. There were no local knights. The other members were three justices of Common Pleas—Wadham, Hankford, and Brenchley—four dukes and two local peers, Thomas Camoys, and Thomas West. It was a bench of magnates, lawyers, and stewards. Clearly the latter were the 'working' justices, appointed in some measure at least for their legal expertise.

It is possible to argue that opportunities for professional administrators were more numerous in the late fourteenth century than they had been a generation or two earlier. The clerks had been displaced from their monopoly of royal and private chanceries, and the monks from theirs of abbey stewardships. Attorneys, moreover, were needed to advise on the complexities of litigation. Perhaps, too, the rewards

[88] Robert Oxenbridge is worth a study in his own right. His descendant Sir Goddard acquired a moiety of the Etchingham estates by his marriage to Elizabeth, daughter and co-heiress of Sir Thomas Etchingham (d. 1482).

[89] Bolney's daughter Agnes married William Gage: thus were laid the foundations of that family's future eminence. Bartholomew's brass is still to be seen in Firle Church. It is illustrated in C. E. D. Davidson-Houston, 'Sussex Monumental Brasses', *SAC* 77 (1936), 175.

[90] Bolney's appointments are summarized conveniently by Searle, p. 422.

[91] *CPR 1396–9*, pp. 99, 231–2.

[92] Ibid. 205.

[93] He acted as a feoffee for the earl and some of his relatives (*CCR 1396–9*, pp. 72, 548).

were greater: cash fees, ready hospitality, and leases on preferential terms were among the many pickings on offer. Yet there was a darker side. The late middle ages were a period of contraction rather than expansion for most great landowners. Fees were being cut back and not only at Battle. It is far from easy in practice to measure the changing extent of opportunity in quantitative terms. Brook, Oxenbridge, and Bolney command our attention because they conform to our image of the successful bureaucrat in the late-medieval 'age of ambition'. But it is doubtful if their like had not been seen before. To go back no further than a couple of generations before Brook, there had been Robert Sharnden, servant of both the abbot of Battle and the archbishop of Canterbury and one-time sheriff and escheator.[94] And a couple of generations further back still, in a different part of the country, there was Robert Thorpe of Longthorpe, steward of Peterborough Abbey and founder of a successful knightly family.[95]

If it is misleading to suggest that professional bureaucrats were a new arrival on the late-fourteenth-century scene, it is no less so to imply that scions of old established knightly houses were no longer capable of doing the jobs which they had once done. Of course they were. And the three lineages with which we began can themselves provide a couple of fascinating studies in the nature of careerism and the rewards that it could bring. Each of the two knights concerned—Sir Andrew Sackville and Sir Robert de Etchingham—rose in the service of secular, not ecclesiastical lordship. If it is worth drawing any distinction at all between their experience and that of the humbler 'professionals' like Brook and Bolney, it is probably this: that the knights were employed chiefly by secular magnates, who valued the chance to strengthen their ties with the leaders of county society, and the 'men on the make' chiefly by ecclesiastical corporations, for whom such political considerations did not weigh so heavily.

Sir Andrew Sackville III was born into a family that was well connected almost in spite of itself. His great-grandfather had been a

[94] The appointment as sheriff may admittedly have been a mistake (or Sharnden declined to accept). Six weeks after it was made, one Reginald Forester was appointed in his place (*PRO List of Sheriffs*, p. 136).

[95] His own career (in the years around 1300) and the rise of his family are discussed by E. King, *Peterborough Abbey, 1086–1310: A Study in the Land Market* (Cambridge, 1973), pp. 50–3, 132–3. It was presumably this Robert who commissioned the remarkable set of early-fourteenth-century wall paintings for his home at Longthorpe, near Peterborough (King, pp. 132–3, and C. Platt, *Medieval England: A Social History and Archaeology* (London, 1978), plate 35).

Montfortian and his grandfather brought up under Edward I's watchful eye and married to one of Queen Eleanor's ladies-in-waiting. His father was married to a Mortimer. And he himself was brought up as a ward of the royal household knight Sir John de la Beche, and married by him to his daughter Joan.[96] The connection with the court that had been forged as a result of his own and his grandfather's minorities was one that was to prove remarkably durable over the generations. When young Andrew prepared for active service in 1336, he did so in the retinue of the king's brother John of Eltham, earl of Cornwall.[97] He seemed destined, given his background and upbringing, to become a knight of either the king himself or one of his kinsmen. But with John's death later in the same year he had to change directions, and by 1340 he had found a new patron in the person of Richard FitzAlan, earl of Arundel. He was accompanying him in Shropshire and the Welsh Marches in 1340 when he was required to appear before the justices of oyer and terminer in Sussex; he did not attend, and his absence cost him an amercement of 20*s*. Twelve months later he obtained cancellation of the fine on the grounds that it was the king's business on which he had been engaged; but the fact that he could claim to have been engaged on it in the company of one so great as Arundel doubtless eased his passage.[98] Arundel, in fact, was by no means the only lord on whose assistance he could call. He may not even have been the first; for it was during roughly this period—the late 1330s—that Andrew made the acquaintance of one of Edward III's closest friends and counsellors, William Montagu, earl of Salisbury. In 1346, two years after the earl's death, when the king was taking depositions in the dispute between Sir John de Molyns and the earl's executors over the ownership of the manor of Stoke Trister (Somerset), he was summoned to give evidence as a confidant of the late earl. He described how he had been asked to deliver letters to Molyns's wife, but denied that he knew anything of the contents, still less of the rights and wrongs of the dispute.[99]

Sackville was wise to be equivocal: better plead ignorance than risk offending a proven thug like Molyns who had recently been restored to favour at court.[100] But if he was indeed unaware of the contents of

[96] See above, p. 12. [97] PRO E101/19/36 m. 1.

[98] He was pardoned on 27 Oct. 1341 (*CPR 1340–3*, p. 336).

[99] *CPR 1345–8*, pp. 139–41.

[100] A thug is what Molyn was: N. M. Fryde, 'A Medieval Robber-Baron: Sir John Molyns of Stoke Poges, Buckinghamshire', in *Medieval Legal Records, edited in memory of C. A. F. Meekings*, ed. R. F. Hunnisett and J. B. Post (London, 1978), pp. 198–223.

his late master's letter, one possible reason could have been pressure of commitments elsewhere. He had taken part in most of the opening engagements of the Hundred Years War. In 1338 he had crossed to the Low Countries with Edward III, and in the following year he had returned there.[101] In the winter of 1340/1 he was attending, as we have seen, on the king's business in Shropshire in the company of the earl of Arundel. Then, five years later, he was abroad again. He contributed to the great victory won by the English at Crecy, and was present for at least the early stages of the siege of Calais;[102] but a premature return to England in late 1346 or early 1347 prevented him from witnessing the final surrender of the town in August.[103]

He had served for long enough, however, to earn a few of the rewards of active service. In September 1346 he received a general pardon, and in the following February a grant of free warren in all his demesne manors.[104] To these expressions of royal generosity he was presumably able to add the ransoms and the other more tangible (but sadly unquantifiable) spoils of war that made campaigning in France so profitable.[105] For all his success on the battlefield, however, he was more than just a seasoned campaigner. He was an energetic administrator whose services were much in demand. It is doubtful if he ever needed to flaunt his qualifications in order to win offers of employment. He simply exploited the growing network of Montagu family connections. In the mid-1340s he was drawn by the marriage of William's elder daughter Elizabeth into the service of her husband, Sir Hugh Despenser, and a few years later by that of William's second daughter Philippa into the service of her own husband, Roger Mortimer, earl of March.[106] After the deaths of William Montagu and

[101] Protections were issued in his favour on 26 Dec. 1338 and 20 Apr. 1339 (*Cal. Treaty Rolls 1337–9*, no. 750; PRO, C76/14 m. 9). He was with the king at Valenciennes in Sept. 1339 (*CPR 1338–40*, p. 395).

[102] *CPR 1345–8*, pp. 497, 529.

[103] On 10 May 1347 he and Sir Gerard Braybrook witnessed a release from one John de Cottele to Edmund and Joan de Malyns, their Oxfordshire neighbours from Henton, near Chinnor (Magdalen College, Oxford, M. Oxon. vii, 133). This attestation must surely have taken place in England—in Oxfordshire—and not in France. I am grateful to Mr John Titterton, who is working on the Malyns family, for drawing this document to my attention.

[104] *CPR 1345–8*, pp. 497, 529.

[105] Was his white surcoat with the ermine cuffs perhaps one of the spoils of war? See below, p. 173.

[106] For Elizabeth's marriage see GEC iv. 273. It was in Sir Hugh's retinue that Andrew fought on the Crecy-Calais expedition (*CPR 1345–8*, p. 497). On 18 Sept. 1347 he is described as Sir Hugh's steward (ibid. 555). For Philippa's marriage to Roger Mortimer see GEC viii. 445.

Hugh Despenser in 1344 and 1349 respectively it was Roger Mortimer who became his main employer.[107] On 20 November 1355, in his capacity as the earl's chief household steward, he is found authorizing receipts for expenses at Calais, whither he had accompanied his master in Edward III's army.[108] Five years later he and his son accompanied the earl again on the expedition which proved to be the latter's last: he died in France in February 1360.[109] By then Sir Andrew was himself an old man, and he put his soldiering years behind him to settle down in retirement to the life of a (still active) country gentleman. He was sheriff of Surrey and Sussex from 1366 to 1368 and an MP in 1365, 1368, and 1369, on the last two occasions with Sir John Waleys of Glynde. At the same time he was putting the finishing touches to his scheme to transmit his inheritance to his son by Joan Burgess. He ended his life by serving himself as well as he had served his patrons.

Andrew Sackville, though employed over the years by three or four families, in a sense served only one, because his later employers were related to the first. Sir Robert de Etchingham, though working in his case for two employers—Edward II and the elder Despenser—again in reality served only one, because the interests of king and favourite were indivisible. As we have seen, Robert was the second son of Sir William de Etchingham III (d. 1294). He acquired his first taste of active service in 1298, when he accompanied his elder brother William IV at the battle of Falkirk.[110] On subsequent campaigns, however, he accompanied the Despensers.[111] It may have been at Falkirk that he

[107] We cannot say how long Andrew remained in the service of the Montagu and Despenser families after the deaths of William and Hugh. It is unlikely that he withdrew immediately. Quite the contrary: the estates still needed to be administered, and officials still needed to be supervised. It is even less likely that he withdrew quickly when, as in each case, the death was followed by a minority. The services of a competent and influential administrator would have been needed all the more to ensure the safe transmission of the inheritance from one generation to the next. As McFarlane noted (discussing the nobility), 'There was a marked contrast between the transitoriness of male lines and the continuity of so many of the great estates.' It was 'continuity of administrative structure plus continuity of personnel that gave to the great estates . . . the same sort of stable tradition and expertise that the royal governmental institutions gave the Crown'. (*Nobility of Later Medieval England*, pp. 136, 140).

[108] PRO, E43/W.S. 415. For the campaign in Artois and Picardy that year see W. Longman, *The History of the Life and Times of Edward the Third* (2 vols., London, 1869), i. 361.

[109] C76/38 m. 10. For this Roger Mortimer see G. Holmes, *The Estates of the Higher Nobility in Fourteenth-Century England* (Cambridge, 1957), p. 17.

[110] *Scotland in 1298*, ed. H. Gough (Paisley, 1888), p. 213.

[111] Robert went with the king in 1301 (PRO, C67/14 m. 2) and with Hugh Despenser

first met them. It is difficult to think where else a young esquire from Sussex could have established contact with a family whose territorial base lay so far from the southern seaboard. They were not yet the power in the realm that they were to become in Edward II's reign, and the influence that they could exert in a county like Sussex would have been decidedly limited. But such considerations were probably of marginal concern to Robert. Being a person of, as yet, little consequence himself, he was looking not so much for the protection of a local magnate as for an opportunity to advance himself in the service of anyone willing to take him on. Such a person he found in the elder Hugh Despenser sometime around 1300. And the connection, once forged, was to prove equally beneficial to both parties. Robert was to serve the family for the best part of a quarter of a century. And as a result of their rise to power under Edward II he gained an entrée into the royal court. There, on 2 July 1321, he was admitted as a knight of the king's household, with a fee of 40 marks per annum to be received out of the issues of the towns of Rye and Winchelsea.[112]

The timing may be significant. In the summer of 1321 the Despensers' fortunes were in temporary eclipse. Their overweening territorial ambitions in South Wales had provoked such a storm of opposition that Edward was forced to bow to the demands for their exile. By August they were gone, albeit only temporarily. So Robert may have been seeking to compensate for the possible loss of one retaining fee by the acquisition of another. If so, he can hardly be blamed. The pursuit of self-interest was a sensible enough policy in those uncertain months; and he was at least consistent in his choice of political allegiances. In November, however, the younger Despenser returned, and full-scale civil war broke out. Whether or not Robert was present on the field of Boroughbridge in March 1322, when the rebels were crushed, we cannot be sure; but the chances are that he was, because we find him accompanying the royal host on its progress northwards later that year—he and the three esquires with him were in receipt of wages from the Wardrobe from 8 August until 22 September.[113]

The victory at Boroughbridge gave Edward and the Despensers a position of dominance no royalist party had enjoyed since the triumph

in 1303 and 1306 (C67/15 m. 9; C67/16 m. 10). He was also in the retinue of Hugh Despenser at the Dunstable tournament in 1309 (Tomkinson, 'Retinues . . .', 76).

[112] *CPR 1317–21*, p. 599.
[113] BL MS Stowe 553, fol. 60^{v}.

at Evesham half a century before. The favourites became the unchallenged arbiters of power, and their retainers and estate staff the executants of their designs and in some cases the instruments of their tyranny.[114] Robert was not one of the most oppressive of these men: no bills denouncing his conduct were submitted in the wake of his master's downfall. But a man in his position, poised between court and county, was bound to gain in both wealth and influence. He was seen by the people of Sussex and Kent as one who could intercede on their behalf with the king and his favourites. It was on his advice, for example, that in January 1322 Edward pardoned William son of Ralph le Mareschal and his brother John for not appearing to answer a charge of assault, and at his request twelve months later that a general pardon was issued to one Henry le Pestour of Cranbrook (Kent).[115] Communities as well as individuals looked to him to advance their interests at court. His election as a knight of the shire for Sussex on three occasions before Boroughbridge and one after it was based in part at least on the calculation that none was better qualified than he to secure redress for the county's grievances. But after the middle of 1322, in fact, his responsibilities in his native county dwindled.[116] More important matters further afield were to beckon him.

War had broken out in Gascony between England and France. In October 1323 an impetuous Gascon lord by the name of Raymond Bernard had burned down a bastide the French king was building at St-Sardos and, adding insult to injury, hanged the French official in charge of it. When news of this outrage, committed by someone at least nominally his subject, reached Edward II in London, he immediately wrote to Charles IV to apologize. However, the French judicial machinery had already started grinding into action; and when Edward unwisely requested yet another postponement of the oath of homage he was due to perform for Gascony, Charles retaliated by confiscating the duchy and sending in his army. The extensive collection of diplomatic documents and newsletters relating to the war

[114] Nigel Saul, 'The Despensers and the Downfall of Edward II', *EHR* xcix (1984), 1–33. The most unpopular seem to have been Sir Ingelram Berenger, Sir Thomas Gobion and Richard Cleet, the bailiff on the Berkshire estates (*Rotuli Parliamentorum*, ii. 380–1, 385, 415–16).

[115] *CPR 1321–4*, pp. 52, 246.

[116] He was appointed to a couple of commissions of oyer and terminer in 1323 and 1326 (*CPR 1321–4*, p. 374; *1324–7*, p. 290). Then in the autumn of 1326 he was of course given local responsibilities in the measures that the government took to forestall a possible landing by Queen Isabella (*CPR 1324–7*, pp. 302, 310).

that Dr Chaplais has brought together allows us to follow the subsequent course of events in remarkable detail and to trace the part played in them by some of the humbler participants, Sir Robert de Etchingham included.

In March 1324 Robert found himself called upon to undertake some of the delicate diplomatic duties occasionally required of a royal household knight. He and a Gascon clerk by the name of Bertrand Ferrand were sent to Gascony with letters from Edward II to Charles asking for a delay in process in the Paris *parlement*.[117] They were instructed that, if they found themselves staying in Gascony for longer than sixty days, they were to claim their expenses from the constable of Bordeaux. Whether Bertrand returned within that time is not clear, but Robert certainly stayed for the duration of hostilities, because he was present at the truce-making at La Reole in September.[118] Next, we hear of him being sent by the earl of Kent with Adam Lymbergh, the constable, to tell Edward II what had been happening in the duchy and to ask him to send reinforcements.[119] Lymbergh, and presumably Etchingham too, left Bordeaux on 18 October. The previous day Sir John Felton, the captain of Saintes, had written to the younger Despenser, who was directing the war from England, telling him to pay close attention to what Etchingham would say, 'for surely you will learn Sir, that I will not recover Agenais except by way of Saintonge'.[120] Early in the New Year he was back in Gascony, bringing Christmas greetings from the king to Arnald Caillau, the captain of Blaye.[121] The main reason for his visit, however, was to secure the Agenais, that part of the Gascon borderland which Felton had talked of recovering the previous autumn. There he was sent with Sir Oliver Ingham to secure the allegiance of the local people by handing out retaining fees.[122]

During one of his periodic return visits to England, in the late autumn of 1324, he had been appointed by mistake to a commission of array in Sussex. His name was soon removed, however, because, as the enrolment of cancellation put it, he was 'otherwise engaged' on the king's business.[123] In September 1325 he was off again to Gascony, this time in the company of young Edward, the king's son, who was sent in

[117] *The War of Saint-Sardos (1323–1325)*, ed. P. Chaplais (Camden 3rd Ser. lxxxvii, 1954), pp. 24–5.
[118] Ibid. 63.
[119] Ibid. 81.
[120] Ibid. 83.
[121] Ibid. 171.
[122] Ibid. 153.
[123] *CPR 1324–7*, p. 82.

the place of his father to do homage before the French king for the duchy.[124] But his association with the royal household was shortly to be brought to an end. In September 1326 Queen Isabella and Mortimer landed in England, and Edward II's regime collapsed. On 7 January parliament met in an atmosphere of crisis. It had little choice but to bow to the demands for the king's deposition. On 13 January and the two days following the nobility of England and the knights and sergeants of the royal household, Robert among them, all rode to the Guildhall and took the oath to the new king and the liberties of London.[125] On 25 January Edward III's reign began.

The new regime was very nearly as selfish as the one which it replaced. Nevertheless it showed no hesitation in making use of the services of those who had aided the Despensers in the years of power, and on 14 July 1327 Robert was appointed a keeper of the peace in Sussex with Sir Edward St John.[126] The latter was another erstwhile supporter of the king who, whatever his past sympathies, had no hesitation in defecting to the queen as soon as he could and even in gathering four bags of Chancery rolls to present to her at Hereford.[127] In view of their similarity of outlook and experience the two keepers could have established a harmonious working relationship; but it was not to be. Robert died in September the same year.[128]

Andrew Sackville was an only child (or only surviving child) who had succeeded his father at the age of ten. Already a well-to-do knight when he took his first stewardship he was scarcely dependent for his livelihood on the income he drew from retaining fees. Robert de Etchingham's position was different. He was a younger son. A year before his death he was to succeed his childless elder brother; but for most of his life he had had to make his own way in the world. This difference in personal circumstances affords the only clue we have to the motivation of the two men. They were both successful careerists. But Andrew became so—as far as we can tell—by preference, and Robert by necessity.

[124] Ibid. 170; N. Denholm-Young, 'Edward of Windsor and Bermondsey Priory', reprinted in his *Collected Papers on Medieval Subjects* (Oxford, 1946), p. 164.

[125] *Calendar of Plea and Memoranda Rolls . . . of the City of London, A.D. 1323–1364*, ed. A. H. Thomas (Cambridge, 1926), pp. 12–13.

[126] *CPR 1327–30*, p. 89.

[127] *CPR 1324–7*, p. 337, concerning which see Saul, 'The Despensers and the Downfall of Edward II', 30.

[128] *CIPM* vii, no. 175. For Robert's manor-house at Glottenham see below, pp. 166–9.

The mental world of these knights, however ill documented, is in fact worth a little further discussion, and for this reason: for most of their active lives they must have thought more in terms of the vertical ties of lordship than the horizontal ones of community. That much we can deduce from what we know of their careers.[129] But were they in that respect typical of their peers? Would the majority of the gentry of east Sussex have thought in the same terms? We would be misguided, of course, if we expected to find many others as well connected or as ambitious as these two seem to have been—Roger and Edward Dallingridge come to mind, but no others.[130] It is difficult enough to identify many more members of our three families who were retained by a single lord, let alone two or three. Sir John Waleys may have been one—he is found a few times in the company of Sir Michael de Poynings.[131] His successor Sir William may have been another—he was a supporter of the Appellants in 1387.[132] For the rest the evidence is negative; for that reason, of course, it is not conclusive, but all the same it is telling. It does not give the impression of a local society organized in or around a magnate affinity. Sussex was no Warwickshire, nor even a Norfolk.[133] In so far, therefore, as vertical relation-

[129] It is possible that Andrew's attitudes changed in the last five or six years of his life when he settled down in Sussex. He was obliged now to consider not the needs of an employer but the interests of a local community. In 1368 or 1369 he and his fellow MP Sir John Waleys took to parliament a petition from the counties of Surrey and Sussex complaining about the difficulty of raising the farm as a result of grants of franchises. The original is PRO, SC8/75/3707.

[130] For the Dallingridges see below, pp. 67–8.

[131] Waleys fought in the retinue of Sir Michael de Poynings on the Crecy-Calais expedition (*Crecy and Calais*, ed. Wrottesley, p. 85), and on one occasion in 1347 accommodated that lord and his immediate entourage overnight at Glynde ('Et in probend' equorum diversorum hominum per unam noctem cum domino Mich' de Ponyng usque Lewys 1 qr 1 bus.' GLY 1072, 'Avena'). But whether this wartime attachment lasted into peacetime, or even extended beyond a single campaign, is far from clear. When he crossed the Channel again, in 1355, it was in the retinue of Roger Mortimer, earl of March (PRO, C76/33 m. 10). Presumably he had been recruited by Sir Andrew Sackville, who was also on that expedition. To complicate matters further, John is found in 1353 attesting an indenture of Chichester Cathedral in the company of the earl of Arundel and his retainers (*Cartulary of the High Church of Chichester*, ed. W. D.Peckham (SRS xlvi, 1942–3), no. 815). So if he thought in vertical terms, it was of lords, not a lord.

[132] Goodman, *The Loyal Conspiracy*, pp. 38, 43.

[133] The government of Warwickshire in the early fifteenth century was under the firm control of the earl of Warwick, Richard Beauchamp. According to Dr Carpenter, the network of connections centred on his affinity was so all-embracing that there was no possibility of any split among the county's gentry (C. Carpenter, 'The Beauchamp Affinity: a Study of Bastard Feudalism at work', *EHR*, xcv (1980), 531). Norfolk and Suffolk before the age of the Howards, the de la Poles and, of course, the Pastons is

ships may be identified by the evidence of magnate affiliations, a good number—the majority, surely—of the knightly families of east Sussex cannot be said to have thought in vertical terms.

Did they think in horizontal ones, then? Did the county community act as a substitute for the hierarchical bonds of service and loyalty? The sentiments and reactions aroused by the county (or any other unit of local government for that matter) are harder to identify and to analyse than those associated with lordship. That a county's knights and freeholders could often come together to speak with one voice need not be doubted. They did so in Sussex, as we have seen, in 1297 when they declined to elect two MPs without the presence of the archbishop, bishops, and earls. The question is whether a periodic gathering of a hundred or 150 people had enough common interests, shared enough beliefs, to constitute a community rather than a mere assembly of individuals. If it did, it might have had the government itself to thank for the fact. Over the previous century or more kings had been asking local knights 'to bear the witness of the shire'. What they sought from them was information, but between information and opinion there was no sharp distinction.[134] To that extent the crown's appetite for factual knowledge may have contributed to the formation of political opinions in a corporate setting. By the late thirteenth century the substance of these opinions—often couched in the form of petitions which the MPs would be charged to present to the king[135]—took the form of limiting the crown's demands and regulating the conduct of its officials. It represented the defence of local interest in the face of increasing royal intervention. But whether that local interest took the form of automatic identity with the county through which it was expressed is by no means clear. The suitors to the shire court may have thought less in terms of an abstract county community than of bundles of rights and privileges.

In so far as county feeling fed on resistance to outside encroachments it was of fairly recent growth and of somewhat uncertain

considered in R. Virgoe, 'East Anglia under Richard II', in *The Reign of Richard II*, ed. F. R. H. Du Boulay and C. M. Barron (London, 1971), pp. 218–41.

[134] J. C. Holt, 'The Prehistory of Parliament', in *The English Parliament in the Middle Ages*, ed. R. G. Davies and J. H. Denton (Manchester, 1981), pp. 5, 22.

[135] J. R. Maddicott, 'Parliament and the Constituencies, 1272–1377', in *The English Parliament in the Middle Ages*, pp. 62–9. Such Sussex petitions are PRO, SC8/4/158 (*Rotuli Parliamentorum*, i, 379), for which see below, p. 58, and PRO, SC8/75/3707, for which see above, n. 129.

strength. In so far as it fed on historic or geographical attachments it might have been able to command a stronger and older loyalty. Here a county like Sussex, which was the direct heir to a Saxon kingdom, benefited from being able to draw on a far deeper well of local sentiment than could one like Leicestershire or Nottinghamshire which was the artificial creation of the tenth-century monarchy. But the historic identity of Sussex can hardly be said to have been matched by a corresponding measure of geographical unity. It lacked, as it still lacks, a focal point, a natural centre within equal reach of all parts of the county. Chichester was the accustomed venue for meetings of the county court; but because it lay so far to the west the sheriff sometimes convened meetings at towns further east such as Shoreham, Horsham, and Lewes. His efforts were little appreciated, however, because suitors never knew where to go. In the 1330s they petitioned the king to appoint commissioners to choose a town agreeable to all concerned where the court might in future be held and writs be directed. Three commissioners were appointed, but the outcome of their deliberations is not known.[136] For all its disadvantages Chichester remained the preferred venue for the next 200 years; and it was not until the reign of Henry VII that a solution to the problem was found in the principle of alternation. In future meetings were to be held alternately at Chichester in the west and Lewes in the east.[137]

The problem of finding an acceptable venue proved so intractable because human endeavour, however well intentioned, could not overcome the realities of geography. The county stretched east–west. But its main lines of communication, through the river valleys, ran north–south. There *were* roads that ran laterally along the coastal plain. The Gough Map in the Bodleian Library has a line beginning at Southampton and passing through Havant, Chichester, Arundel, and Bramber, Lewes, Winchelsea, and Rye to Canterbury; and in the Roman age there was a road running a little to the south–east of this linking Glynde with Pevensey which might have survived into medieval times.[138] It was rather when a journey to or from the Weald was contemplated that the problems began. Here the terrain was less tamed, and the conditions therefore were more difficult. For a

[136] The petition is PRO, SC8/4/158, and the commission *CPR 1334–8*, p. 289. This problem was not unique to Sussex. It cropped up too in Cornwall and Kent, for example (R. C. Palmer, *The County Courts of Medieval England, 1150–1350* (Princeton, 1982), pp. 8–13).

[137] 19 Henry VII, c. 29.

[138] R. A. Pelham, 'Studies in the Historical Geography of Medieval Sussex', *SAC* 72 (1931), 182; I. D. Margary, 'Roman Roads from Pevensey', *SAC* 80 (1939), 29–61.

gentleman on horseback it meant a slower, more hazardous journey; for a carter or carrier it meant substantially higher costs. When the sheriff of Sussex was levying prises in 1319, the cost of transporting 300 qrs. of corn from Mayfield to the coast was 5*s*. 8*d*. a league for the Wealden section of the journey as far as Lewes and only 3*s*. 6*d*. for the Downland section from there to Shoreham.[139]

Shoreham was only one of a number of coastal towns from which prises were dispatched to the English forces on the Scottish border. Arundel, Chichester, Seaford, and Winchelsea were likewise authorized to serve as collecting points for their respective hinterlands. Roughly speaking, these hinterlands corresponded to the six divisions of the county into the rapes of Chichester, Arundel, Bramber, Lewes, Pevensey, and Hastings. These units, mid-way between the levels of shire and hundred, bear an obvious affinity to the lathes of Kent; whether they were as old, however, is by no means clear. The earliest division of the county was simply into two halves, east, and west. Its further division into rapes came later and in stages. The three eastern rapes were based on the castles of Lewes, Pevensey, and Hastings, which William the Conqueror had given to William de Warenne, Robert of Mortain, and Robert count of Eu respectively. The whole of the western half of the county King William had given to Roger of Montgomery. A few years later an additional rape was formed for William de Braose with its centre at his castle of Bramber. It was only in the middle of the thirteenth century that the final rape, of Chichester, was formed, probably as a consequence of the partition of the estates of Hugh d'Aubigny, the last earl of Arundel of his line.[140]

The administration of a rape was a fusion of royal and private jurisdictions. The sheriff, who headed it, was usually a minister or tenant of the lord of the rape. Its court, superficially the equivalent of the shire court, had far more of the character of an honorial court. In the fifteenth century, as in the twelfth, the court of the rape of Hastings—to take but one example—met every three weeks at the villages of Sedlescombe and Netherfield alternately. There were no hundred courts—only those of the rape or lathe. The suitors, who were the tenants holding by knight service, were required to hear 'all pleas of life and member' and to report on 'all things which may happen within the hundreds or barony'.[141]

[139] Pelham, 'Studies in the Historical Geography of Medieval Sussex', 170.
[140] L. F. Salzman, 'The Rapes of Sussex', *SAC* 72 (1931), 29; *VCH Sussex*, iv. 1–2.
[141] Searle, *Lordship and Community*, pp. 201–4.

It may be possible to argue, therefore, that in these south-eastern counties the overlap of lordship with local decentralization combined to make the rape or lathe, and not the county, the principal focus of local activity; and local activity may in turn have bred local sentiment.[142] But the nature and depth of that sentiment cannot be gauged; and there is no easy way of saying how long it lasted. It is possible that it waned fastest in Pevensey and Hastings, where from the late twelfth century, if not before, the lords were absentee. On the other hand, the very absence of a strong vertical tie may have compelled the local gentry to look to each other for more mutual support and protection. The strong impression given of the community in the rape of Pevensey by the roll of indictments of Queen Philippa's justices of oyer and terminer of 1352 is of a society self-sufficient, accustomed to running its own affairs and brooking no outside interference.[143] There is no reason to suppose that it was any different a quarter of a century later when Sir Edward Dallingridge and his allies took issue with the novelty and intrusiveness of the lordship of the queen's successor, John of Gaunt.[144]

If this analysis bears any resemblance to reality, the familiar picture of a county community may have to be discarded, in this case at least, in favour of that of a county of communities. The Sussex élite was made up of groups of men who lived sometimes at considerable distance from each other and who belonged to quite separate networks of clientage and collective responsibility. The composition of these groups can be reconstructed in outline by looking at, for example, marriage connections, witness lists to charters, and the personnel of magnate affinities. It is an exercise which has to be done for a generation at a time. Society was never static: families died out in the male line; affinities dissolved and reconstituted themselves around new leaders. But the areas of geographical recruitment seem to have remained roughly the same throughout. If we take the handful of families from whom the principal local office-holders were selected in the reign of Richard II we find that six came from east and three from

[142] It is perhaps worth mentioning in this connection that commissions of array were sometimes appointed not for the whole county of Sussex but individual rapes or groups of hundreds within rapes (*CPR 1374–7*, p. 500; *1385–9*, p. 387). Commissions for the upkeep of walls and ditches ('de walliis et fossatis') were also, of course, appointed for specific stretches of coastline—usually in this area, the Sussex-Kent borderlands (*CPR 1381–5*, p. 134).

[143] PRO, JUST 1/941A.

[144] Walker, 'Lancaster v. Dallingridge'.

west of the Ouse.[145] To some extent the retinue of the earl of Arundel provided a focal point. All three of the west Sussex office-holders were, understandably enough, dependants of his, and so too were Dallingridge and Sackville from the east.[146] But there the overlap ends. The knights and esquires of east Sussex had few other ties with their peers from the western half of the county. The explanation is straightforward: they had few if any lands there. Sir William de Etchingham IV was an exception in that he held the manors of Stopham, Linch, and Yapton, brought to him by his marriage to Eve de Stopham; but in the next generation these were all sold. The minimal importance attached to them by the Etchingham family could hardly have been made any plainer.

If our three families were landowners in only one part of their own county, they did nevertheless hold property outside it, the Sackvilles in Oxfordshire and East Anglia, the Etchinghams and Waleyses in Kent, and the marriage alliances they each contracted reflected this wider distribution of their interests. The Sackvilles, for example, contracted two alliances, widely separated in time, with the family of Malyns, who were near neighbours of theirs in Oxfordshire. Sir William de Etchingham VI married the widow of Sir William de Brien of Seal in Kent, and his sister and aunt married respectively Sir Arnald Savage and Sir William Brenchley, both of them Kentish knights. Sir Richard Waleys II married the daughter of one of his own tenants from the manor of Newenden, just over the county boundary into Kent.[147] An analysis of the marriage ties contracted by these and other families resident in the same area goes a long way to reinforce the conclusion we have already reached—that the gentry of eastern Sussex had more in common with their neighbours east of the river Rother than they did with their peers in the west of their own county.[148]

In criticism of this conclusion it may fairly be objected that many of the sources we have used illuminate not a man's friends but only his acquaintances: the fellow members of an affinity, for example, would not necessarily have been the partners with whom he chose to while away his leisure hours. This point is a valid one; but it should not be

[145] John Ashburnham, Robert Etchingham, William Waleys, Thomas Sackville, Edward Dallingridge, and John St Clair from east of the Ouse, and Edmund Fitz-Herbert, Edward St John, and William Percy from west.

[146] See above, p. 36.

[147] For sources see above, Ch. I, *passim*.

[148] Again, see Ch. I, *passim*.

pressed too far. The division between public and private, between hours spent on business during the day and hours spent in private in the evening, would have meant far less in the middle ages than it does today. Most lords—even making allowance for the desire for greater privacy evidenced by the architectural developments of the period—would have spent most of their waking hours in public view. Likewise, pursuits like hunting which today we might consider leisure activities could easily assume the character of public, even political, acts.[149] Similar considerations apply to the interpretation of marital connections. Certainly, at this level of society, partners were generally chosen not by the young people themselves but by their elders in accordance with political or dynastic imperatives.[150] Yet it has to be said that preachers expected love or affection—'dilectio socialis', as Guibertus called it—to develop within the conjugal relationship.[151] And sometimes it did: Joan Etchingham settled down so comfortably in the household of her Kentish husband that when she came to make her will many years later she forgot all about her relatives back home.[152]

If friendship is to be found anywhere, it is to be found surely in the choice of a man's feoffees. Friendship implies trust. And feoffees were people in whom trust had to be placed. But they were also people of whom a measure of expertise was expected. Lawyers would therefore have to be included, however open to question their own honesty may have been. Retainers then usually made up the remainder of a magnate's feoffees, and fellow gentlemen a knight's or esquire's. The men to whom Sir Reginald Cobham of Sterborough (Kent) granted his estates in June 1404 were a fairly typical group of feoffees: they included three fellow landowners (and probably friends)—Sir Thomas Sackville, John Hadresham, and John Culpeper, two estate officials—Alan St Just and John Ingram, and two lawyers—Sir William Brenchley and John Brook.[153] Sir William Etchingham VI's feoffees

[149] Like Dallingridge, Sackville, and Medsted hunting in Ashdown Chase (Walker, 'Lancaster v. Dallingridge').

[150] For an exception, see below, pp. 179.

[151] D. L. d'Avray and M. Tausche, 'Marriage Sermons in *ad status* collections of the Central Middle Ages', *Archives d'Histoire Doctrinale et Litteraire du Moyen Age* (1980), 114.

[152] The will is printed by G. O. Bellewes, 'The last Savages of Bobbing', *Arch. Cant.* xxix (1911), 167. There are no bequests to friends or relatives in Sussex.

[153] *CCR 1405–9*, p. 469. It is not surprising that Sackville should have been one of the feoffees. His residence of Buckhurst was only about seven or eight miles, as the crow flies, from Sterborough. Alan St Just was also an official of Sir Philip St Clair of Brambletye, another neighbour (account roll of Heighton St Clair, 1403/4: SAS G1/46). John Hadresham is commemorated by a brass in Lingfield Church, where the Cobhams

reveal a similar social and occupational profile: Sir John Pelham and Sir Arnald Savage were up-and-coming knights well connected with the Lancastrian court, Richard Wakehurst and Robert Oxenbridge local stewards probably with legal experience, John Chidecroft, Robert Castrete, and John Halle local esquires, perhaps again lawyers.[154]

The feoffees whom Sir Andrew Sackville empowered to transmit his inheritance to his illegitimate son included no such distinguished company—no knights or fellow gentlemen at all. Apart from a single lawyer they were principally estate officials: John Spicer, his receiver and later his steward, Thomas Preston, probably another steward, Peter de Hoo, rector of Alfriston, and William Halden, the recorder of London.[155] The omission must surely be significant. In view of the uncertainties likely to attend the succession of a bastard—and one, moreover, who was under age—Sir Andrew chose to appoint not fellow landowners who might have been tempted to take advantage of his misfortune but employees and officials with a vested interest in the continuance of the Sackville line.

The possibility that he might not have been willing to trust his knightly neighbours may appear to be at variance with the earlier observation that the gentry were bound by mutual obligations of trust and co-operation. But the enterprise on which this man had embarked was surely exceptional. And in view of the smoothness with which it was in the event to be implemented, he may have underestimated the degree of goodwill felt by his peers and neighbours. But his choice of feoffees raises other questions. What did he think of Spicer, Halden, and the others? Did he see them as equals or inferiors, friends or employees? Can a legal arrangement of this sort honestly be taken as a guide to the kind of company that a man kept? In the case of someone as well connected as Sir Andrew there is a natural temptation to suppose that he mixed more easily with barons and bannerets than he did with his own staff—until, that is, we look at the men with whom the fourth Sir William de Etchingham chose to pass his time. Here was another rich, well connected knight, a man in his later years

of Sterborough were themselves buried. It is illustrated by M. Stephenson, *A List of Monumental Brasses in Surrey* (reprinted Bath, 1970), p. 344.

[154] BL Add. MS 39375 (Edwin Dunkin quoting PRO, CP40/615 m. 404d, a document now deemed unfit for production).

[155] BL Add. MS 39375, fol. 85^{v}. For Spicer see SAS CH 255, 257, and for Halden see below, p. 184.

summoned to parliament as a lord. He knew the earl of Surrey; and he had attended the funeral of Lady Tregoz.[156] There is no denying his standing in society. Yet the men with whom he went hunting in 1308/9 were not these grander sort but his tenants and immediate neighbours Alan and Thomas Buxhill.[157] Alan was the father of the more famous Sir Alan Buxhill who was to achieve notoriety in 1378 for his part in the murder in the precincts of Westminster Abbey of two esquires, Robert Hauley and John Schakell.[158] Until his name was sullied by this breach of sanctuary he had enjoyed a fine reputation as a soldier, and in 1372 had been honoured with election to the Garter.[159] He died in 1381 a rich man. But his origins had by no means been so exalted. His father had held the estate of Bugsell, in Salehurst, from Sir William de Etchingham, in addition to which he was granted by his lord the manors of Socknersh (Sussex) and Bryanston (Dorset).[160] Although the income from these manors could scarcely have been princely—Bugsell was said to have been worth £7. 14*s*. 0*d*. in 1325[161]—he assumed the rank of knight and saw active service in the Scottish wars, on one occasion in the company of Sir Andrew Sackville II.[162] But it was Etchingham who was his main friend and patron. He is found associated with him in grants to Robertsbridge Abbey.[163] He received from him a livery of 7 qr. of wheat in 1309/10; and his brother, with his falconer, received a payment of 4*s*. in the same year.[164]

With the passage of time the links between the Etchinghams and the Buxhills weakened. The younger Alan won a position in the royal household, and there he mixed with a new circle of friends.[165] The

[156] *Lewes Cartulary*, i. 55; *CPR 1313–17*, p. 653; GLY 996 (Beddingham account roll, 1309/10, 'Avena').

[157] GLY 996 (Beddingham account roll, 1308/9, 'Avena').

[158] A. H. Cooke, *The Early History of Mapledurham* (Oxford, 1925), p. 35.

[159] There is an account of his career in *Dictionary of National Biography*, sub Buxhill.

[160] The licence for the alienation of the former was granted in July 1314 (*CPR 1313–17*, p. 161). For Socknersh, which is in the parish of Brightling, see *VCH Sussex*, ix. 229. The draft conveyance is AMS 5789, no. 16.

[161] *CCR 1323–7*, p. 429.

[162] PRO, C67/16 m. 5.

[163] For grants of his to Robertsbridge Abbey, both witnessed by members of the Etchingham family, see *Calendar of Charters and Documents relating to the Abbey of Robertsbridge* (London, 1873), nos. 288, 298, and for grants by the Etchinghams witnessed by him, ibid., nos. 300, 321. No. 326 is witnessed by John and Thomas de Buxhill. Alan and Sir William de Etchingham are associated in a recognizance, *CCR 1313–18*, p. 618.

[164] GLY 996 (Beddingham account roll, 1309/10, 'Frumentum' and 'Expense forinc'').

[165] He had become a knight of the king's household by 1365 (PRO, E101/396/2 fol. 56r). But his links with Sussex did not disappear completely. In that same year, 1365, he

relationship between the Etchinghams and another local family, the Ores, was on the other hand to be more enduring. The Ores were lords of the manors of Ore and Guestling, north-east of Hastings.[166] In the early-fourteenth century young Wiliam de Ore was serving as a page in the Etchingham household when his father died, and his overlord, Humphrey de Bohun, earl of Hereford, had to initiate a suit in Common Pleas to enforce custody of his person.[167] William's son John (d. 1361) was one of Sir James's nominees to investigate the affair of the 'Knellesflote' sluice in 1348, and he acted as an attorney or feoffee for him and for other members of the family.[168] John's second son and eventual heir Robert also acted as an attorney and was associated with Sir William V in many commissions in the 1370s.[169] Holding as they did two medium-sized manors, the Ores were better endowed than the early Buxhills. They were middling gentry; but they were certainly not in the same bracket as the Etchinghams or the Sackvilles.

Another close acquaintance—one hesitates to use the word friend—of William IV was a man by the name of Peter de Ros. It is doubtful if he was a scion of the Yorkshire baronial house of that name; for a young man of such distinguished birth, even one entrusted to the wardship of another lord, would surely never have been sent out, as Peter and an unnamed colleague were in 1309–10 to audit the accounts of two of Sir William's manors.[170] He was a local man, probably—an estate official and yet more than an estate official. He bears a strong resemblance to those protean administrators often found in the

was at Burwash to witness a deed alongside Sir William de Etchingham, Sir John St Clair, William Tauk and John and Roger Ashburnham (*CCR 1364–8*, p. 178).

[166] For the descent of the manor of Ore see *VCH Sussex*, ix. 87. David Martin tells me that the estate at Guestling was overlooked by the *VCH*.

[167] William de Ore crops us as a 'garcio' in the Beddingham accounts for 1307–11 (GLY 996). For the suit by the earl of Hereford see Dunkin's notes (BL Add. MS 39373, quoting PRO, CP40/162 m. 174). I think this William must be the man commemorated by the Purbeck slab recently discovered lying face down in the ruins of the chancel of the old church at Ore. Its inscription, as reconstructed by Father J. Bertram, begins: '+ Villame de Ore signor de Ore gist ici . . .' The date is early fourteenth century. Now William the 'garcio' of 1307–11, is still alive in 1332 (*VCH Sussex*, ix. 87), but dead, it seems, by 1340, when his successor—his son, one supposes—is active (e.g. *CPR 1338–40*, p. 559). The approximate date of William's demise would therefore fit the approximate date of the slab. For the latter I have relied on Father Bertram's manuscript *Sussex Brasses and Slabs, Part II: East Sussex*, ii. p. 89.

[168] *CPR 1348–50*, p. 80; PRO, JUST 1/941A m. 32.

[169] *CPR 1364–67*, p. 13; *1374–77*, p. 410; *CCR 1377–81*, pp. 91–2.

[170] GLY 996, Beddingham account roll, 1309–10, 'Expense forinc'.

households of the gentry who could turn their hand equally to land management, man management, and soldiering. His skills were unspecialized, and his tasks varied.[171] He might be called upon to tour the estates one month and to set out for Scotland the next: in July 1298 he was to be found fighting alongside his lord at Falkirk.[172] Such sudden changes in role strike us today as incongruous; but they were probably far from unusual for a man in service.[173]

Ros probably spent most of his life in and around the household. His landholdings (if he had any) cannot now be identified. The Ores and the Buxhills, on the other hand, definitely did have a foothold on the ladder of landowning society—the former a higher one than the latter, though strangely it was the latter who in two successive generations assumed knighthood. Neither family could rank with the Etchinghams in terms of wealth of esteem. The ties that each had with them derived from tenure or geographical proximity rather than social parity. But in so far as the marks of favour which they received from the Etchinghams are any guide, they may be said to have been among the closer acquaintances of the heads of that family.

In reality gentle society was a good deal less stratified and a good deal more fluid than appearances might lead us to suppose. The bonds of tenure and locality that brought men together could count—in the early fourteenth century, at least—for more than those of economic and social differentiation which kept them apart. Moreover, it was a surprisingly dynamic society. This is not the place to consider how different the Sussex of 1400 was from the Sussex of 1300. Some of the changes, indeed, scarcely need pointing out—those in magnate leadership, for example. The FitzAlans had succeeded to the place once occupied by the Warennes; and the focus of lordship in the county had moved westwards to Arundel. But it may be that other, less obvious differences counted for more. Our concentration on the affairs of the three families of Etchingham, Sackville, and Waleys, which kept going in the male line throughout the fourteenth century, has blinded us to the undoubted changes that took place in the composition of the county élite in the same period. The names of

[171] In addition to his estate duties he is also found acting as a feoffee or deforciant (*Cal. Robertsbridge Charters*, no. 291; *Feet of Fines for the County of Sussex, from 1 Edward II to 24 Henry VII*, ed. L. F. Salzman (SRS, xxiii, 1916), no. 1268).

[172] *Scotland in 1298*, ed. H. Gough (Paisley, 1888), p. 213.

[173] Where staff were relatively few, as they were on the gentry estates, they had to be prepared to switch jobs. Clough makes this point in her discussion of the staff of Sir John Pelham ('Estates of the Pelham Family', p. 192).

Aldham, de la Lynde, Harengaud, and Wardedieu, prominent in 1300, had gone by a century later.[174] These extinctions can be reckoned as part of the cellular process by which one family was always being replaced by another. Certainly, there was no great shrinkage in the overall numbers of the Sussex knightly families in this period, because only the estates of the Aldhams disappeared into the hands of an already established knightly lineage.[175] The rest passed to families which were on the early stages of the ascent into landed society. The Harengaud estates were inherited by one Richard Charles, of whom relatively little is known.[176] The de la Lynde and Wardedieu estates were inherited by the brothers Roger and Edward Dallingridge respectively—men of whom considerably more is known, and who were to figure a good deal more prominently in Sussex history than had the lineages they replaced.[177] Sir Roger was a companion-in-arms of Sir Michael de Poynings and a retainer of the earl of Arundel.[178] Like Sir Andrew Sackville, his near contemporary, he was both a soldier and an administrator, and the experience he gained as steward of the dowager countess Warenne was to serve him well later in life when he settled down to assume the burdens of office-holding that came the way of so many retired veterans. He was appointed sheriff of Surrey and Sussex for the year 1371/2 and was many times elected an MP.[179] Edward Dallingridge initially followed in his brother's

[174] Sir Francis de Aldham, who was executed in 1322, for his complicity in the Contrariants' rebellion, was succeeded by his nephew Sir John St Clair (G. Ward, 'The Aldhams', *Arch. Cant.* xl (1928), 31–4). For the Harengauds see *VCH Sussex*, ix. 187. The exact dates of death of the last members of the Harengaud, de la Lynde and Wardedieu families are not known.

[175] The St Clairs: on whom see W. Budgen, 'Excete and its Parish Church', *SAC* 58 (1916), 143–5. Their main residence was Brambletye, near East Grinstead.

[176] He, anyway, is the person who succeeded Sir Thomas Harengaud at Icklesham in roughly the 1360s (*VCH Sussex*, ix. 187).

[177] Part of the explanation for the relative lack of interest in Sussex affairs shown by the de la Lynde and Wardedieu families—though Sir Henry Wardedieu was an MP for Sussex in October 1302—must be that they held lands outside the county, the former in Somerset, the latter in Northamptonshire. When the Wardedieus were in Sussex, they resided at their manor of Bodiam, where they had a substantial moated homestead, described by W. Douglas Simpson in *SAC* lxxii (1931), 69–99.

[178] Roger Dallingridge fought in the retinues of either Sir Thomas or Sir Michael de Poynings in 1338/9, 1340, 1345, 1346 (PRO, E36/203 fol. 126ᵛ, C76/15 m. 21, C76/20 m. 16, and *Crecy and Calais*, ed. Wrottesley, p. 137). He was also a feoffee of the earl of Arundel (*CPR 1364–7*, pp. 237–8). For evidence of military service I am, as usual, grateful to Andrew Ayton. I shall keep footnote annotation on the Dallingridge to a minimum as I propose to deal with them in more detail on another occasion.

[179] He was five times a knight of the shire between 1360 and 1377. He was appointed a

footsteps: he joined the Arundel retinue and won his spurs on the battlefields of France. But in later life, either because he was the younger son or because he was more able or more ambitious than his brother, he chose to strike out in the direction of national rather than local politics.[180] All the same, he acquired a stake of his own in Sussex society by marrying the heiress of the Wardedieu family; and by the time he had added to her lands the ones he inherited from his brother he could pass himself off convincingly enough as the champion of local autonomy against the intrusive lordship of an outsider like John of Gaunt.[181] Bodiam Castle survives to attest his greatness in local society.

Ten years before Dallingridge was granted his licence to crenellate, Scotney Castle, a few miles to the north-west in the valley of the river Bewl, had been rebuilt by its owner, another younger son—Roger Ashburnham. John, his father (d. 1335), was lord of the village of that name near Battle. They were not a rich family—a single manor hardly constituted an ample patrimony—and the eldest son was to be troubled by indebtedness in the 1370s.[182] Roger, the younger one, however, was luckier. From his mother, Isabel, widow of John de Grovehurst of Horsmonden (Kent), he inherited a couple of properties along the Kent border, at one of which—Scotney—he chose to settle.[183] But if he was better endowed than his elder brother, he can hardly be said to have enjoyed the level of landed income that would have enabled him to finance building works on the scale of those he undertook in the 1370s. He rebuilt Scotney from the foundations. To do that he must have had other sources of income to call upon. But whether these represented the profits of war or the profits of service we cannot say. We have no evidence to guide us beyond that of the

commissioner of array in 1367 (*CPR 1364–67*, p. 365) and a justice of the peace in 1375 (*CPR 1374–77*, pp. 136–7).

[180] He was a member of the commission of inquiry into the state of the royal household appointed in 1380. Thereafter his responsibilities were mainly local until in 1390 he became a member of the king's council (J. F. Baldwin, *The King's Council in England during the Middle Ages* (Oxford, 1913), pp. 132–3). In 1392 he was appointed warden of the City of London after the arrest of the mayor and sheriffs (C. M. Barron, 'Richard II and London', in *The Reign of Richard II*, p. 184).

[181] Walker, 'Lancaster *v*. Dallingridge'.

[182] This seems to be the implication of his lease of the manor of Ashburnham to two clerks, William Steel and Nicholas Despagne, in 1371 (*CCR 1369–74*, pp. 295–6). According to *VCH Sussex*, ix. 127, he was later imprisoned for a debt to one Nicholas Greve of £600.

[183] *Scotney Castle* (National Trust, 1979), p. 5.

building itself. What survives is fragmentary; but it is enough to leave us in no doubt that, like other younger sons whose careers we have observed, Roger Ashburnham did well, perhaps even very well, for himself.[184]

It is surely no coincidence that the two castles raised in east Sussex in this period were both built by successful careerists. They served not only to deter the French but to set the seal on the rise to power of their owners. Yet power once won might prove difficult to retain. The careerist, having jumped on fortune's wheel as it ascended, might find himself cast off when it turned downwards. Certainly it seemed to be spinning faster than ever at the end of the fourteenth century. In 1397 Richard II began, as Walsingham put it, to 'tyrannize' his people.[185] The earl of Arundel, one of the Appellant lords of a decade earlier, was arrested and executed, and his Sussex estates divided between Thomas Mowbray, earl of Nottingham, and John Holland, earl of Huntingdon.[186] Both were court-based magnates, and Holland for one was a complete stranger to Sussex.[187] Without the burden of loyalty and expectation that went to constitute a magnate's standing in a locality, he could have done little in practice to advance Richard's cause, and the appointment as sheriff in 1398 of the pro-Ricardian Sir William Fiennes may well have done more than the territorial redistribution to tilt the local balance of power in the king's direction.[188] Given time, Mowbray, Holland and their allies could have

[184] There is not even a licence to crenellate to enable us to date the rebuilding of Scotney with any precision. Sir Roger died some time in the 1390s. Christopher Whittick has drawn my attention to BAT 2711, which appears to be a memorandum of *c.*1440 embodying submissions by rival claimants to the manor of Ewhurst, which Sir Roger had acquired. One of the claimants, Thomas Ashburnham, his great-nephew, gives us a precious insight into his last will. Roger, he said, had conveyed the manor to four feoffees—Richard Alard, Thomas Newenton, Thomas Remys, and his own father John Ashburnham—giving them instructions to enfeoff first his widow and then, after her death, his son William; in the event of the latter's death before 21, they were to give John Ashburnham the option of buying it for £100. William, he said, predeceased his mother. John Ashburnham then died, leaving his son Thomas, the claimant, to try to exercise his option to buy it on the death of Roger's long-lived widow. By that time, he said, he had been cheated of it by the remaining feoffees.

[185] *Annales Ricardi Secundi*, in *J. de Trokelowe et H. de Blandeforde . . . Chronica*, ed. H. T. Riley (Rolls Series, 28, 1866), p. 199.

[186] *CPR 1396–99*, pp. 175, 209–10, 280–1.

[187] Mowbray, however, held the lordship of Bramber. His principal retainer in the area was Robert de Halsham of West Grinstead (PRO, KB27/356 m. 43).

[188] The appointments to the shrievalty in November 1397 and in 1398 were of a political nature in most counties. Richard appointed either knights of the royal household or retainers of magnates, like the Hollands, whom he had promoted. Fiennes was not a household knight, and I cannot establish any connection with Holland or

built up their position; but time was not given to them.[189] In 1399 Richard was toppled, and Bolingbroke mounted the throne as Henry IV. In January 1400 Holland joined the conspiracy against him and, after its exposure and collapse, himself fell victim to the executioner's axe. Ten months later Arundel's attainer was reversed and his son Thomas restored in blood and in honours.[190]

To a background of such dizzying political instability the Dallingridges may be judged, with the gift of hindsight, to have played their cards rather well. Edward's son John had joined the Lancastrian retinue in the early 1390s, about five years before his father's death.[191] The decision must have been made in the light of Gaunt's local position as lord of the rape of Pevensey, but its wisdom was confirmed by the events of 1399 which made Sir John, as he now was, a household knight of the new king with an annuity of 100 marks for life.[192] He had emerged richer and more secure from the change of dynasty; but for all his loyalty to the house of Lancaster he never became the power in the land that his neighbour Sir John Pelham was to be. Pelham's is the most remarkable case-study in the rewards of careerism that late medieval Sussex can provide. In the space of a single lifetime he laid the foundations of what a couple of centuries and a few good

Mowbray. On the other hand, his presence on Richard's expedition to Ireland in 1394–5, which was very much the household in arms, points to a connection with the court and its activities. William is commemorated by a splendid brass at Herstmonceux, illustrated by C. E. D. Davidson-Houston, 'Sussex Monumental Brasses', *SAC* 78 (1937), 88. John Salerne's appointment that previous year, in the immediate aftermath of the arrest of the erstwhile Appellants, is also an interesting one. He was a resident of Rye and a past mayor of the town (*CCR 1385–9*, pp. 290, 659), but had never before been sheriff. Was he leader of the pro-Ricardian faction in Rye? In addition to granting Holland the estates of the earldom of Arundel, Richard helped to build up his position a little further by granting him the reversion of the wardship of Sir John Arundel's manor of Old Shoreham (*CPR 1396–99*, p. 198).

[189] Holland had started recruiting in Sussex by the time of his master's fall, because the two servants of his in whose fates the author of the *Traison et Mort de Richard II* showed particular interest both bore Sussex names—Sir Thomas Shelley, his master of the household, and Hugh Cade, his chief butler (J. J. N. Palmer, 'The Authorship, Date and Historical Value of the French Chronicles on the Lancastrian Revolution: I', *Bull. John Rylands Library*, 61 (1978–9), 165).

[190] GEC i. 246.

[191] He accompanied Henry of Derby on his expedition to Prussia (*Expeditions to Prussia and the Holy Land made by Henry, Earl of Derby*, ed. L. Toulmin Smith (Camden NS lii, 1894), 120).

[192] *CPR 1399–1401*, p. 69. He surrendered this allowance when in 1405 he received instead the custody of the castle and lordship of Bramber, forfeited by Thomas Mowbray (*CPR 1405–8*, p. 26).

marriages later was to be a great ducal fortune. Yet he himself was the heir to no more than a humble dynasty of coroners. They had pretensions to gentility, certainly, but lacked the means to support it. In 1348, 1349, 1356, 1366, 1369, and 1372 writs were issued for the removal from office of his father on the grounds that he was insufficiently qualified.[193] For the son, however, magnate service was to be the path to success. He joined the Lancastrian affinity in the early 1390s, at about the same time as John Dallingridge, and in 1394 or 1395 was appointed constable of Pevensey by John of Gaunt.[194] He was still holding that office, and evidently considered a man of complete dependability, in the summer of 1399, when Bolingbroke landed there briefly before sailing north to Ravenspur, where the Lancastrian presence was thicker on the ground.[195] Pelham probably sailed with him, leaving his wife to conduct a spirited defence of Pevensey against the assaults of Richard's partians in Sussex.[196] He was knighted at Henry's coronation on 13 October 1399 and later became a member of the Council and finally in 1412 treasurer of the realm.[197]

John Pelham was the principal beneficiary in east Sussex of the revolution that had put Henry IV on the throne. His proximity to the fount of patronage enabled him to acquire the jurisdictional power he needed in order to establish control over his own 'country'. In 1400 he received a grant of the rape as well as of the castle of Pevensey in tail male, and ten years later of the reversion of the rape of Hastings.[198] He benefited too from that liberality bred of necessity which was so often

[193] *CCR 1346–49*, p. 451; *1349–54*, p. 1; *1354–60*, p. 260; *1364–68*, p. 219; *1369–74*, p. 371. In 'The Estates of the Pelham Family', p. 28, Dr Clough argues convincingly that Sir John was the son of Thomas II, the coroner whose replacement was provided for in these writs. She makes him the son in turn of Thomas I, a coroner from 1313 to 1320. But it is clear from a sixteenth-century transcript of some Battle Abbey court rolls now in the Huntington Library (microfilm in ESRO, XA 3/4, para 307) that Thomas II was not the son but the grandson of Thomas I. *His* father John was the son of Thomas I. When Thomas I died in 1327, his son had already predeceased him, and his grandson Thomas II, a minor, was said to be his heir. The future Sir John Pelham was born in about 1365.

[194] The known facts of his career are conveniently assembled in J. S. Roskell, *The Commons in the Parliament of 1422* (Manchester, 1954), pp. 208–11. For his appointment as constable of Pevensey see R. Somerville, *History of the Duchy of Lancaster*, i (London, 1953), p. 380. For his lawlessness see R. L. Storey, 'Liveries and Commissions of the Peace', in *The Reign of Richard II*, ed. Du Boulay and Barron, pp. 134–5.

[195] *CPR 1396–99*, p. 596; M. McKisack, *The Fourteenth Century* (Oxford, 1959), p. 492.

[196] The famous letter that she wrote to her husband on 25 July 1399, one of the earliest in English, is printed in M. A. Lower, *Historical and Genealogical Notices of the Pelham Family* (privately printed, 1873), p. 11.

[197] Roskell, *The Commons in the Parliament of 1422*, pp. 209–10.

[198] Lower, *Historical and Genealogical Notices . . .*, pp. 11–12.

the subject of complaint in Henry IV's parliaments. The custodies of the lordship of Bosham and of the temporalities of the see of Winchester, the keepership of the New Forest, the marriage of Roger Fiennes—these and other grants made him a richer man, and provided him with the means to become richer still through the purchase of manors.[199] According to a valuation of his lands made in 1403, his income from rents alone came to £870.[200]

A certain likeness is to be observed between his career and that of his contemporary from Cheshire, the first Sir John Stanley.[201] Stanley showed perhaps the greater flair for the exercise of power. As a former servant of Richard II he was required to effect a transfer of allegiance that Pelham never had to make. But both men used their positions at court to construct in their respective localities a power base which was to be the foundation for future family greatness.[202] A certain likeness is to be observed, too, between Pelham's career and that of Sir Edward Dallingridge. Each man united in his person the role of royal lieutenant and leader of local society. Oddly enough, it was the absence, or relative absence, of magnate lordship in east Sussex after 1347 that made the rise of such men both possible and necessary. Each in his own generation acted as a power broker regulating the flow of royal patronage and reconciling the demands of the crown to the rhythms of local life: articulating the grievances of the local community and harnessing the resources of the crown to advance himself at the expense of his rivals. It was the presence of these able, ambitious and successful men that, more than anything else, made the Sussex of 1400 a different society from the Sussex of 1300.

[199] For grants and custodies see *CPR 1399–1401*, p. 143; *1401–5*, pp. 49, 326, 329, 497; *CFR 1399–1405*, pp. 240, 314. On 20 Dec. 1411, after he had become Treasurer, he issued a pardon to himself of all debts and arrears due at the Exchequer (*CPR 1408–13*, p. 369). The burdens of office have their reward.

[200] Roskell, p. 209.

[201] As described by M. J. Bennett, *Community, Class and Careerism*, pp. 216–19.

[202] The Stanleys became earls of Derby, and branches of the Pelhams earls of Chichester and dukes of Newcastle.

III

THE COMMON LAW

It may or may not be a tribute to the effectiveness of magnate management in the county that Sussex saw so little, if any, of the violent feuding between rival lords or their retainers that disturbed the peace of other counties in the late middle ages.[1] The last earl Warenne, it is true, was engaged in open warfare with Thomas of Lancaster by the late 1310s, but it was only his Yorkshire estates that suffered as a consequence.[2] Surrey and Sussex, where Lancaster lacked any following, remained peaceful. Indeed, in central Sussex John de Warenne was himself the only lord with a following of any size, and through alliances with lesser lords like Michael and Thomas de Poynings he was able to reconcile most of the local gentry to his rule. Of those who remained outside the fold the only one to cause him any trouble was Sir John Waleys of Glynde; and even he knew when to keep his peace.

Waleys was accused of harbouring a nest of malefactors who kidnapped local people and held them to ransom.[3] The sheriff of Sussex, it was later said, ignored the problem for fear of incurring his hostility. On 13 August 1338, therefore, Sir William de Warenne, acting, so he claimed, under the authority of a commission issued to his father, the earl, and the other overseers of the keepers of the peace, seized Waleys at Glynde and abducted him to Lewes where they held him against his will for the next eight weeks.[4]

[1] The best stories from the fifteenth century are told by R. L. Storey, *The End of the House of Lancaster* (London, 1966). For some fourteenth-century examples see Phillips, *Aymer de Valence*, pp. 261–5, and J. G. Bellamy, *Crime and Public Order in England in the Later Middle Ages* (London, 1973), Ch. II.

[2] Maddicott, *Thomas of Lancaster*, pp. 207–8.

[3] The case has recently been printed in *Select Cases of Trespass from the King's Courts, 1307–99*, i, ed. M. S. Arnold (Selden Soc., 100, 1984), 44–5. My own account has been taken from the pleadings of Hilary Term 1350 (PRO, KB27/358 m. 16).

[4] By the time that the case reached court, more than a decade after the events being described, memories had evidently faded. John alleged that the abduction took place on the Tuesday after the Feast of St Laurence, 10 Edward III, that is 13 Aug. 1336. William in reply claimed as his authority a commission to the earls of Surrey, Arundel, and Huntingdon, John Mowbray and John Hampton, dated 1 Aug. 1338. This was the appointment of overseers of keepers of the peace (*CPR 1338–40*, p. 141). Since there

After his release, oddly enough, Waleys did nothing. It was not until shortly after the death of the earl nine years later that he at last brought actions of trespass against his captors, principally Sir William de Warenne but also other family retainers like Sir John Bigod and Sir John de St Pier.[5] The reason for the delay must surely have been that he feared to bring the actions during the earl's lifetime or, at the very least, supposed that his chances of securing a favourable verdict would be negligible. Once the threat of magnate interference was removed, however, he felt free to proceed. In Easter 1348 the suits were entered, and formal demands (or exigents) made in the county court for the appearance of the accused. After five such demands had been made in vain, they were all outlawed; but a sixth demand, made in May 1349, prompted Sir John Bigod to obtain a pardon from the king, on production of which in King's Bench he was immediately released.[6] Only the case against Warenne went to a jury: a verdict of guilty was returned, and John awarded damages of £40.[7] He had scored a limited but notable victory over the late earl's party. Yet he never seems to have felt for the earl himself the hostility he reserved for his retainers. On hearing of his death in the summer of 1347 he had no hesitation, as we have seen, in joining the other local knights who had returned from France to attend the funeral.[8] This was not the action of a man filled with hatred, however sorely he nursed his wounds.

In the second half of the century the pace of change in Sussex quickened. The Arundels succeeded to the place once occupied by the Warennes. They themselves were temporarily displaced after the

was agreement that John's abduction took place just a fortnight later, it seems likely that 1336 was a slip for 1338. William de Warenne was an illegitimate son of the earl (Fairbank, 'The last Earl of Warenne and Surrey', pp. 243, 248, 254). In 1346 the earl referred to him as 'our dearest son' (*Lewes Cartulary*, i. 67).

[5] The suits were entered in the Michaelmas Term immediately following the earl's death (PRO, KB27/350 m. 49d). Sir John was the son of Sir Ralph, for whom see above, p. 32 (*CCR 1330–3*, p. 528). In 1353, following Warenne's conviction, Sir John de St Pier obtained a pardon for his own part in the seizure of Waleys (*CPR 1350–4*, p. 430).

[6] PRO, KB27/357 Rex m. 30. Why was a sixth exaction made? The outlawries had after all been pronounced after the customary five exactions had been read out.

[7] PRO, KB27/368 m. 57.

[8] See above, p. 35. The nature of John's original offence remains unclear. He was said to have harboured a nest of malefactors who ran something like a modern protection racket. Such allegations are fairly common in the plea rolls. They may mean exactly what they say. On the other hand, they may just be formulaic. It is possible that John had challenged the earl's authority more directly than we can now tell. Could he, a tenant of the archbishop and a resident of the rape of Pevensey, have been interfering in the earl's own rape of Lewes?

execution of the second Earl Richard in 1397. And John of Gaunt, as we have seen, drove deeper inroads into the autonomy of the east Sussex gentry community than had any outsider before him. Yet it is doubtful if these disturbances contributed to any noticeable rise in large-scale lawlessness. Or, if they did, it never found its way into the plea rolls. Just one violent ruction stands out: the feud already noticed between John of Gaunt and Sir Edward Dallingridge and his two accomplices Sir Thomas Sackville and Sir Philip Medsted. All three were clients of the earl of Arundel, whose temporary eclipse at court in 1384 allowed Gaunt to proceed against them by both special bill and jury of indictment. The charges, which amounted to 'a campaign of systematic intimidation against the duke's estates and officials',[9] included firing the duke's underwood and driving off his livestock, preventing his steward from holding his lord's court at Hungry Hatch and depriving him of his court rolls, and finally murdering William Mouse his sub-forester. But in essence it was a dispute about lordship, in particular about lordship over men. The crux of the matter was Dallingridge's opposition to Lancaster's court at Hungry Hatch which, he alleged, was drawing suitors away from his own hundredal court of Dean (i.e. Danehill, in Horsted Keynes).[10] Legally he was on poor ground: the court was by no means the novelty he claimed it to be, and Gaunt had no difficulty in securing a conviction. But politically he was stronger. He was articulating the grievances of a closely knit gentry community which disliked the intrusive (and arguably aggressive) nature of Gaunt's lordship.[11] Moreover, with the aid of his patron, the earl of Arundel, he secured release from prison little more than a forthnight after his committal to custody. In the long run, therefore, Gaunt pulled back. He had established the principle that he was able to hold a court at Hungry Hatch; but in practice he abandoned it. When the local gentry closed ranks, there was little that he or anyone else could do.[12]

If Sussex avoided the worst excesses of late-medieval magnate feuding, it was also spared the bold but disruptive activities of the outlaw gangs. The Folvilles of Leicestershire, the Coterels of

[9] Walker, 'Lancaster *v.* Dallingridge', 88.

[10] Ibid. 92–3. Hungry Hatch survives as a house on the lane into Fletching from Piltdown. Fletching was the very heart of Dallingridge country: no wonder Edward objected.

[11] Ibid.

[12] Or, as Simon Walker puts it, 'against a tight-knit gentry community of this kind, even the greatest of English magnates could not act until he was sure of his ground' (Ibid., 89).

Nottinghamshire, the Musards and Kingscots of Gloucestershire may have contributed by expert example to the disorder of the times;[13] but they found no direct imitators in Sussex in the fourteenth century, despite the ideal cover afforded by the Wealden forests.[14] Nor could the county field any really villainous knightly malefactor to match Buckinghamshire's John de Molyns or, at a more modest level, Gloucestershire's James Clifford.[15] It had its corrupt office-holders, of course—which county did not? In the 1352 'trailbaston' proceedings Sir Thomas Hoo, sheriff from 1349 to 1351, was accused of extorting goods and chattels from the queen's tenants in the hundred of Netherfield in the name of raising prises.[16] But this kind of malpractice was ubiquitous. Sir John Waleys of Glynde (if the accusations against him are to be believed) is the only knight in east Sussex whom we can present as conforming to the stereotype of the late-medieval 'well-born bandit'.

The stereotype may therefore be more in the nature of a caricature. The well documented cases of a few notorious malefactors may have served to deflect attention from the good behaviour of the rest. Perhaps they have. But good behaviour in the middle ages was always relative. The fact that a knight was never the subject of an indictment means not that he never engaged in wrong doing—only that, if he did, he was never brought to book. Activities like poaching and trespassing, arm-twisting and labouring of juries, indeed all the acts of thuggery and extortion that were part and parcel of everyday life in landed society, would as likely as not have gone unreported and unpunished unless King's Bench or a general 'oyer and terminer' commission happened to visit the county. And in Sussex such visits were far fewer than they were elsewhere.[17]

[13] See respectively E. L. G. Stones, 'The Folvilles of Ashby Folville, Leicestershire, and their Associates in Crime, 1326–1347', *TRHS* 5th Ser. 7 (1957), 117–36; Bellamy, *Crime and Public Order*, Ch. III; R. H. Hilton, *A Medieval Society: the West Midlands at the End of the Thirteenth Century* (London, 1966), pp. 248–61; Saul, *Knights and Esquires*, pp. 178–9.

[14] It may simply be, of course, that they were never brought to book. But if a gang had been operating in 1352, its activities would surely have been brought to the attention of the justices of oyer and terminer who visited the rape of Pevensey that year.

[15] N. Fryde, 'A medieval robber baron: Sir John Molyns of Stoke Poges', in *Medieval Legal Records*, pp. 198–221; Saul, *Knights and Esquires*, p. 176.

[16] PRO, JUST 1/941A m. 26.

[17] PRO, JUST 1/941A, though classified with the assize rolls, is in fact a commission of oyer and terminer. There were no King's Bench visitations of Sussex in the fourteenth century. On the other hand it was visited by the justices of assize every fifteen or twenty years when they made their regular circuits of all the southern counties.

We can agree, then, that a good deal of petty lawlessness probably never came to light. But we can also agree that, if a particularly heinous crime were committed, such as the murder of young Edmund de Pashley in 1328, the king and his justices would soon have heard about it. To that extent the relative absence from the membranes of the plea rolls of cases of serious disorder in Sussex may be taken to be a genuine reflection of the relative stability of the county's society. Such a conclusion would certainly accord with what we have already learned of the cohesiveness of that society. The absence of rival and overlapping spheres of magnate influence helped to deliver fourteenth-century Sussex from the kind of fate that overtook Norfolk in the age of the Pastons or Devon in the age of the Courtenays and Bonvilles.[18] Furthermore, the existence in the eastern half of the county of a closely knit local gentry community facilitated the settlement locally of many disputes which in a more individualistic society might have been solved by recourse to 'self-help'.[19]

The relative absence of large-scale gentry violence need not therefore imply that there were any fewer sources of tension in Sussex than elsewhere, only that they were resolved more peacefully. A low level of lawlessness might be quite compatible with a high level of litigiousness: indeed, it might well be a consequence of suitors' very satisfaction with the workings of the courts of common law. Such satisfaction cannot, of course, be measured in the quantitative terms beloved of modern historians; but usage can, and in so far as it can be interpreted as an index of satisfaction,[20] its message is clear. In the fourteenth century litigants of all classes and backgrounds had no hesitation in initiating actions in the courts. It has been calculated that in the early 1330s Common Pleas, the busiest of the central courts, was dealing with roughly 8,500 cases a year:[21] relatively few of these actions were determined by judgement of course, but all the same the case-load was heavy, indeed very heavy. Norfolk and Yorkshire provided the largest number of actions—roughly 1,000 each—and Sussex, a less populous

[18] The Pastons are their own witness (*Paston Letters and Papers of the Fifteenth Century*, ed. N. Davis (2 vols. Oxford, 1971, 1976). For Devon see M. Cherry, 'The Struggle for Power in Mid-Fifteenth Century Devonshire', in *Patronage, The Crown and the Provinces in Late Medieval England*, ed. R. A. Griffiths (Gloucester, 1981), pp. 123–44.

[19] Fifteenth-century Derbyshire is a possible example of an 'individualistic' society: S. M. Wright, *The Derbyshire Gentry in the Fifteenth Century*, p. 119.

[20] A big 'if'. It may simply be an index of necessity.

[21] R. C. Palmer, *The Whilton Dispute, 1264–1380* (Princeton, 1984), p. 6. The scale of my debt to this immensely useful book will become clear as the argument progresses.

county, some 200 per annum: about the national average or slightly below.

Still more interesting than the number of actions is the number of people bringing them. Dr Palmer has recently estimated that nearly 15,300 different persons were caught up each year as litigants in the processes of Common Pleas.[22] This remarkable statistic places beyond all doubt the high level of involvement with the king's courts that existed in England in the middle ages. The point can be demonstrated more strongly still if the figures for a single county are examined. According to Dr Palmer, the 630 cases originating from Bedfordshire in the years 1331–4 involved no fewer than 1,379 different individuals—and this in a county which would have had difficulty mustering more than a couple of dozen resident knights. Clearly, plaintiffs and defendants were drawn from the ranks of all the free, not just the nobility and manorial lords.[23]

Litigation, of course, was an expensive business, and few could afford to indulge in it regularly.[24] Dr Palmer estimates that in the three-year period from 1331 to 1334 only 10 per cent of the litigants from his two sample counties of Bedfordshire and Cornwall were involved in more than a couple of cases. These men—and occasionally women[25]—it may surely be assumed, were the richest proprietors: the ones with wide landed interests that needed to be protected. Yet even within this small élite wide variations are noticeable in propensity to litigate. In our own county of Sussex, for example, the Etchinghams stand out as regular plaintiffs and their neighbours, the Waleyses, by contrast, as only occasional ones.

The figures presented in Tables I–III are based on a fairly thorough perusal of the rolls of the courts of King's Bench, Common Pleas and of the justices of assize. But they relate only to the county of Sussex. Had the search been extended to include other counties of southern England the heavy preponderance of the Etchinghams, the majority of whose manors lay within a few miles of their home, might have appeared less marked than it is; and the Sackvilles, who held manors in Oxfordshire, Essex, and Suffolk, might have appeared relatively more litigious than they in fact do. But what cannot be explained (or explained away) by juggling with the figures is the inactivity in the

[22] Ibid. 8.

[23] Ibid. 6–7.

[24] See below, pp. 93–4.

[25] Only single women or widows, that is. Married women could not sue or be sued in their own right.

Table I The Litigation of the Etchinghams 1300–1412

Action	Number initiated by the family	Number initiated against the family	Number between members of the family
Trespass	43	4	
Novel disseisin	1	6	2
Quare impedit	1		
Account	2		
Collusion		1	
Dower		1	1
Custody		1	
Warranty	1		
Right	1		
Covenant	1	1	
Intestacy			1*
Debt	1		
Formedon	1		
Statute of Labourers	1		

* Action presumably brought by executors or feoffees.

Table II The Litigation of the Waleys family 1300–1418

Action	Number initiated by the family	Number initiated against the family	Number between members of the family
Trespass	10*	2	1
Debt	2		
Dower			1
Waste			1
Custody		1	
Extortion		1	
Detinue	1		

* This figure includes three separate suits initiated by Sir John against those who seized him in 1338.

Table III The Litigation of the Sackvilles 1300–1410

Action	Number initiated by the family	Number initiated against the family	Number between members of the family
Trespass	4	6	

courts of the Waleys family—and what activity there was was confined largely to the lifetime of Sir John. He was the most active of his line in county politics and likewise the most active in the courts, both as plaintiff and defendant. His father and uncle had made equally little impact in either. Their personal reluctance to become involved in what today would be called public life must be counted as one reason for their absence from the courts of common law; but it was by no means the only one. What did as much to influence the level and location of the family's litigation was their position as tenants of the archbishops of Canterbury. It was to the archiepiscopal court that they owed suit, and in that court that differences with fellow tenants would have been settled.[26]

Collectively the heads of our three families were plaintiffs far more often than they were defendants. It is clear, then, that if they were typical of their peers, the gentry could not have been litigating principally with each other, or a rough balance would have been struck. An examination of the plea rolls suggests that the great majority of the individuals against whom they brought suits were social inferiors: reeves in actions of account, villeins in prosecutions under the Statute of Labourers and free tenants and other small proprietors in actions of trespass. Trespass was undoubtedly the plea most commonly brought in King's Bench in the fourteenth century. It may not have been so in Common Pleas. Professor Nielson showed that of the 6,000 cases entered on a typical roll of the 1330s by far the most numerous, totalling roughly 1,500, were actions of account. Next came actions of debt, totalling just under 1,000, and then detinue and trespass accounting for roughly 500 each.[27] If pleas in King's Bench are brought into the picture, however, there can surely be little doubt that over the century as a whole the knightly class made far greater use of trespass than of any other action.

In origin a trespass was simply a wrong. It was a personal, not a real, action: concerned, that is to say, not with the restitution of rights but with making amends for past transgressions. The great majority of such wrongs were of little concern to the king, and by the early thirteenth century it was settled that the Chancery clerks should only issue writs for those trespasses done 'with force and arms against the

[26] For the archbishop's 'principal courts' of Canterbury and South Malling see Du Boulay, *Lordship of Canterbury*, pp. 299–300.

[27] N. Neilson, 'The Court of Common Pleas', in *The English Government at Work, 1327–1336* (3 vols., Cambridge, Mass., 1940–50), iii. 273–4.

king's peace' ('vi et armis et contra pacem regis'). Three principal kinds of wrongdoing were encompassed within this idea of forcible trespass: the breaking of a close, the taking and carrying away of goods, and assault and battery.[28] But there is no need to take the descriptions of this nature found in the plea rolls at face value. They were the straitjackets into which the actuality had to be fitted to win recognition from the royal courts. In 1466, for example, Henry Hull brought an action against one Richard Orynge alleging that the latter 'with force and arms, that is, with swords, bows and arrows, broke his close and did by walking with his feet tread down and consume his grass and commit other enormities' said to include straying beasts and accidental shootings. Yet the Year Books show that all that Orynge did was enter the plaintiff's land to pick up the clippings from his thorn hedge.[29]

Clearly, if a plaintiff wanted to bring a case into King's Bench or Common Pleas, he would find a way of doing so. But trespass cases were not always so lacking in substance. When Sir William de Etchingham brought no fewer than nine separate writs against groups of malefactors whom he alleged to have trespassed on his property on 13 June 1381, he was seeking redress for what must have been an assault involving as many as 50 or 100 people.[30] The date on which it is said to have taken place is a clue to its possible significance. Thursday 13 June 1381 was the day on which the Commons of Kent and Essex entered London. The Kentishmen had assembled at Blackheath on the previous evening, having covered the seventy miles from Canterbury in two days or less. Rochester and Maidstone had already fallen to them. Sir William's estates all lay close enough to the border with Kent to feel the backwash of this turmoil. The trespassers he named in the suit, though obviously not members of the main rebel army, were men of similar background to those who were, and included in their number a substantial minority with non-agricultural occupations: John Landys carpenter, Simon Standon carpenter, John Shortwode bootmaker, Walter Geffray shoemaker, Robert Harry tailor.[31] Only a few of these men seem to have been Sir William's own tenants. John

[28] J. H. Baker, *Introduction to English Legal History* (London, 2nd edn., 1979), p. 57.

[29] The marvellous example quoted by A. Harding, *The Law Courts of Medieval England* (London, 1973), p. 111.

[30] The suits were entered *en bloc* in Easter Term 1383 (PRO, KB27/488 m. 47).

[31] For the social composition of the rebel bands see R. H. Hilton, *Bond Men Made Free* (London, 1973), pp. 176–95.

Balard of Udimore obviously was one; perhaps there were others whom we cannot now identify. But far more came from adjacent villages owned by other lords such as Dallington, Brightling, Burwash, Bodiam, Lamberhurst, and even Hastings. Clearly the attack on Sir William's property cannot be explained in terms of his unpopularity with his own tenants. Nor can it be dismissed as but one of a number of indiscriminate attacks on the property of the land-owning class. It was not. The only other lord in east Sussex who seems to have suffered in a similar way, and at the hands of roughly the same group of men, was Sir Roger Ashburnham.[32] What made these two men the objects of rebel discontent was their role in the assessment and collection of the hated poll taxes. Ashburnham had been a collector in Sussex of the graduated poll tax of 1379.[33] Etchingham had been a collector of both of the two earlier poll taxes of 1377 and 1379, and was appointed again in November 1380 to collect the very tax which was to spark off the revolt.[34] No wonder that after this experience he declined to be appointed a collector ever again.[35]

The outcome of Sir William's suit is not known. After a couple of years it disappeared from the rolls, as did most trespass suits, without judgement having been given. So many medieval lawsuits ended in this way that the conclusion must be drawn that proof by jury was not the principal object sought by most plaintiffs. Indeed, in the case of a trespass suit there was little to be gained by proceeding thus far. Being a personal, not a real, action it brought only an award of damages not restitution of property. And the damages in any case were never very substantial—the sum of £10 awarded to Sir William Waleys in 1391 was slightly higher than average.[36] They were often scarcely enough to cover the costs involved in bringing the suit in the first place.

It is difficult to identify the circumstances that led the Etchinghams and the other families to initiate so many trespass actions in the two central courts. In many of the cases no pleading ever followed the

[32] PRO, KB27/500 m. 15.

[33] *CFR 1377–83*, p. 143.

[34] *CFR 1377–83*, pp. 57, 143, 187, 225.

[35] He obtained an exemption from office-holding on 5 May 1381 (*CPR 1377–81*, p. 624), and he had to invoke it in December the following year to secure cancellation of a fresh appointment as a tax-collector made a month previously (*CFR 1377–83*, pp. 340, 344). The clear primacy in the causes of the Peasants' Revolt of the levying of the poll taxes is a point made by Nicholas Brooks, 'The Organization and Achievement of the Peasants of Kent and Essex in 1381', in *Studies in Medieval History Presented to R. H. C. Davis* (London, 1985).

[36] PRO, CP40/523 m. 123d.

entry of the suit and, as we have seen, when it did it is not necessarily a reliable guide to what happened. But where sufficient information is supplied an apparently trivial dispute over, for example, the straying of some beasts is often found to be the key that unlocks a larger dispute over the ownership of a tenement. In 1400, for example, Sir William de Etchingham alleged that on 14 May 1399 Peter Church of Ticehurst, his son Richard and his brother John took wood to the value of £5 from his property at Hawkhurst (Kent). Peter denied the charge, and claimed that the wood lay on a tenement which he leased from Alice, widow of Sir Edward St John. Sir William, in reply, gave a completely different account. He claimed that he had bought the tenement from one Stephen Eridge who in his turn had been enfeoffed by Stephen Betenham who, he said, had been the owner long before the alleged lease was supposed to have been granted.[37] The differences between the two sides had been reduced to a single point of fact on which a jury could be asked to decide. An order was made for a jury to be assembled, and the case was adjourned. But, come the next term, the jury did not appear, and no judgement was ever given.

Arguments of a similar nature were heard a couple of years later in Common Pleas when Sir William Waleys brought a trespass suit against his son John and his accomplices John Warere, John Isenhurst, John Sygnet, and Thomas Bost. He alleged that they had broken his close at Mayfield and had cut down 100 trees there to the value of £20. Sir William claimed that the close in question constituted the vill of Bainden, which he had granted for his own lifetime to his half-brother Richard. Afterwards, he said, he had granted Bainden and Hawksden, which adjoined it, to his son John, and he had sent one John de Croxton to deliver seisin to him.[38] This he duly did, but he omitted to ensure, as he should have done, that Richard, who remained in possession at Bainden, attorned, like the other free tenants, to John. Because of this oversight, claimed Sir William, John's seisin was incomplete. For his own part John replied that his father had granted Richard a pension of 10 marks assigned on Bainden without conceding to him an interest in the tenement itself: he was himself the tenant of both Bainden and Hawksden as a result of the delivery of seisin made by John de Croxton. The case was adjourned, and no judgement was given.

[37] PRO, KB27/555 m. 30.

[38] In the case of Bainden, the grant must have been of the *reversion* of the manor, since Richard was then occupying it.

Detinue cases opened up a similar Pandora's box of tangled descents and disputed rights. In 1299 Sir William de Etchingham found himself summoned in Common Pleas to show why he was detaining two bullocks which he had taken from the abbot of Bayham's property at Kechenham. Sir William argued that the carucate at Kechenham was held of him by the abbot for the service of one-sixth of a knight's fee, and that he had distrained the beasts because the abbot had not performed the homage that was due on his own succession.[39]

In English law real and personal actions were sharply distinguished from each other. Yet, as Dr Palmer has shown, the distinction was by no means so clear in practice as it was in theory. It is true that an allegation of trespass represented a claim against a person, not a claim to a thing; but when the injury committed involved the breaking of a close or the taking of chattels, a personal action could be very much part and parcel of a dispute about the ownership of land.[40] Not all trespass actions raised such broader issues, of course. Many amounted to no more than the ordinary acts of hunting and poaching which were an everyday occurrence in the demesnes and parks of the upper classes. Many more amounted to nothing more serious than the incident over the collection of the hedgeclippings. But a fair proportion—exactly how many it is impossible to say—were acts that raised, and sometimes were intended to raise, questions of right or title. Trespass may seem a strange way of raising these questions, an even stranger way of settling them. But in fact late-medieval litigation was a game played according to an elaborate and recognized set of conventions. No litigant ever dreamed of settling an issue by immediate resort to writ of right. To do so risked using a sledge-hammer to crack a nut. Rather he would begin with a less awesome action and then proceed to successively higher ones until his opponent's bluff was called. The first step was to initiate an action of trespass. The wrongdoer then knew that his opponent meant business. He could choose to settle at that stage; or he could let the issue go before a jury. If he did, and he lost, he could choose fresh ground on which to fight his case—by bringing an assize, perhaps. Dr Palmer has aptly described this system of resolving disputes as 'a tiered system of litigation'.[41]

[39] PRO, CP40/127 m. 110. For Kechenham, which is close to Etchingham, see *VCH Sussex*, ix. 214.

[40] Palmer, *The Whilton Dispute*, pp. 175–6.

[41] Ibid. 212.

Undeniably there was much for the courts to argue about. Land was a resource keenly competed for. It brought wealth and status and, in the case of manorial lordships, power over men. In the centuries that followed the Norman territorial settlement increasing social differentiation and the growth of the land market had encouraged the diffusion of property rights among an ever larger number of proprietors. Every manor was punctured, like the proverbial sponge, by pockets of free tenements and by allodial lands that were on the way to becoming sub-manors. In an age when rights were defined more often by custom than by written record, a challenge resulting in a legal action was often the only way of establishing what the tenurial position actually was. Small wonder that there was so much litigation.

Litigiousness, however, as we have already stressed, need not necessarily have bred violence. Large-scale and long-running feuds between noble and gentle families, like the famous one involving the Courtenays and the Bonvilles, were the exception rather than the rule. Fourteenth-century Sussex can show not a single such example. Co-operation between the leading magnates and the clannishness (as it may appear) of the local gentry oligarchies together had the effect of curbing the level of violence employed by parties to a suit. Many of the actions brought by knightly proprietors—undoubtedly the majority of those brought by the Etchinghams—were anyway not against their peers, but against their inferiors. And these are unlikely to have been accompanied by the massing of rival gangs. Where regulation and self-control seem to have been at their weakest, oddly enough, was in the settlement of disputes within the family. These could occasionally be every bit as virulent as feuds between competing magnate affinities.

Undoubtedly the fiercest seen in Sussex in the middle ages was that which beset the Pashleys after 1327. Its origins lay in the confused matrimonial history of the Exchequer Baron, Sir Edmund de Pashley, and the dynastic ambitions of the woman with whom he lived in his later years.[42] Sir Edmund had died in March 1327.[43] A year later his thirteen-year-old son, Edmund junior, was murdered at Coulsdon (Surrey). The coroner's inquest had no difficulty in establishing that the boy's assailants were Adam Saule and Edward le Peleter, acting on instruction from Margaret de Basing who also gave them shelter afterwards. Saule was arrested in London and held at Newgate pending

[42] The following paragraphs summarize my article in *History Today*, 34 (Aug. 1984), 30–5. As notes were not given there, I shall supply them here.

[43] *CIPM*, vii, no. 32.

trial. Margaret in the meantime was understood to be living at her manor of Pashley whither the under-sheriff was dispatched to arrest her. But when he and his men arrived, they were set upon by a gang led by Margaret's de Basing relatives and forced to flee in fear of their lives.[44]

Margaret claimed to be Sir Edmund's widow. Yet at the very time when she was resisting arrest in Sussex, proceedings for dower were being initiated in Common Pleas by someone called Joan who likewise claimed to be his widow. However, she ran into difficulties and decided to petition the king. She could not, she alleged, obtain possession of the dower lands to which she was entitled because they were occupied by Margaret, widow of William de Basing, with whom Sir Edmund had been living for some time, though instructed by his confessors to return to his lawful wife. Margaret saw that Edmund would never marry her, but she wanted her sons by him to inherit the Pashley estates. So she had him poisoned first, and then his eldest son William. Yet she could not murder his youngest son Edmund 'without shedding Christian blood', so she had him murdered at Coulsdon. These crimes, Joan concluded, were common knowledge in the counties around Sussex. She therefore pleaded redress.[45]

This petition may, of course, be a tissue of lies; but it is unlikely. It makes statements which can be supported from other sources. The young boy and his valet *were* murdered at Coulsdon. Sir Edmund, his father, *had* died less than a year before, and his eldest son William *had* disappeared from the scene at about the same time. It is possible, then, that Margaret did, as Joan alleged, have them poisoned. Moreover, there is one other piece of evidence which has a bearing on the problem. Joan, as we have seen, backed her claim by suing in Common Pleas for recovery of the dower lands she said Margaret was detaining. According to an entry on the plea roll, Joan maintained that Edmund had dowered her at the church door of St Mary Magdalen, Old Fish Street, when he had espoused her. The bishop of London testified to the truth of this, and Joan was adjuged to have proved her case.[46]

Two possible interpretations may be considered. Either Sir Edmund had been betrothed to Joan and no more, leaving him free to

[44] Pashley Manor lies a mile or two south-east of Ticehurst. The present, very attractive, house dates from the sixteenth and eighteenth centuries, but to the rear are earthworks which may be those of an earlier moated house on the site. My wife and I are very grateful to the owners, Mr and Mrs Sellick, for showing us round in 1983.

[45] PRO, SC8/266/13293.

[46] PRO, CP40/273 m. 1d.

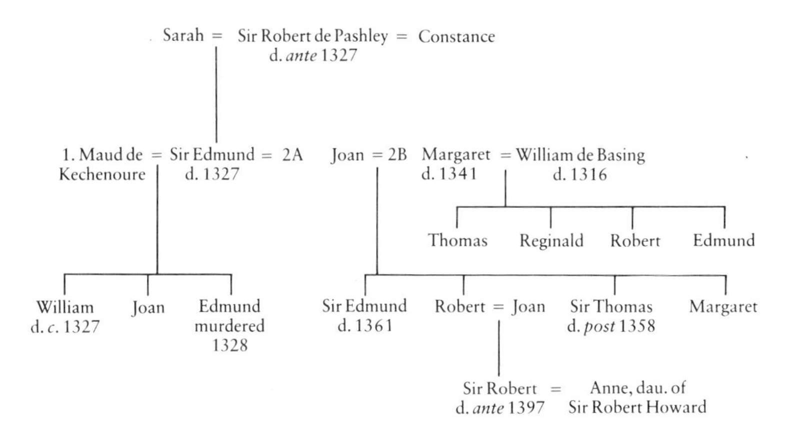

The Pashley Family

marry Margaret. Or, if Joan's petition is taken at face value, Edmund had indeed married Joan, but later left her to live with Margaret. This at least can be said for the latter interpretation, that if it is accepted then everything else falls into place. Whatever may have been the legal status of his relationship with Margaret, there can be no doubt that Edmund spent the rest of his life with her. She bore him four more children—Edmund, Thomas, Robert, and Margaret—and it is only natural that she should have wanted them to succeed to the Pashley estates. The trouble was that the three sons by Edmund's earlier marriage stood in the way. So she planned to dispose of them one by one. William disappeared in about 1327, and Edmund junior, as we know, was murdered. That left John. He must have been beyond Margaret's reach, for there is no suggestion in Joan's petition that she was able to get rid of him. Quite the contrary: it was he who was to carry on the fight against Margaret for the next ten years or more.[47]

If the exotic mix of bigamy, thuggery, and murder found in the Pashley case is impossible to parallel elsewhere, bitter family quarrels between rival heirs to an inheritance were nevertheless a relative commonplace in landed society: quarrels in which litigation in the courtroom might well have been accompanied by the use of some arm-twisting outside it. It is difficult to imagine, for example, that a man so quarrelsome as Simon de Etchingham would have taken kindly to the

[47] For John's raids on the properties of his opponents see PRO, KB27/281 m. 43d, and *CPR 1334–8*, p. 205.

attempt that his elder brother Robert made in the 1320s to deprive him of his inheritance. Simon was a difficult man; but for once he seems to have had right on his side. William de Cessingham, acting as attorney for Robert's widow, virtually made his case for him. He admitted that Robert had granted his estates to feoffees and had received them back entailed on himself and his wife and the heirs of their bodies with remainders successively to his nephews James, John, and Richard, the sons of Richard. Simon by implication was disinherited. Had he stood only a distant chance of inheriting, he might have been prepared to acquiesce. But, Robert being childless, he was in fact the heir at common law, and only this settlement stood between him and the succession. In 1328 therefore, after Robert had died—as expected, without issue—he brought an assize of novel disseisin against his widow. The jury agreed that he should recover seisin, and he duly took possession of the estates.[48]

Later generations of the Etchingham family, so far as we can tell, lived in harmony—or comparative harmony, allowing for the occasional dower case.[49] At the end of the fourteenth century it was the turn of the Waleyses to fall out. Sir William had provided for his younger half-brother, as we have seen, by granting him an annuity of 10 marks which he later converted into an interest for the donor's lifetime in the sub-manor of Bainden, in Mayfield. Subsequently, however, he had granted Bainden—or, rather (one assumes) the reversion of Bainden—and the adjoining manor of Hawksden to his son John, and he arranged for a servant by the name of John de Croxton to deliver seisin to him. Croxton accordingly went to Bainden, and delivered seisin to the new tenant but, by an extraordinary oversight, omitted to ensure that Richard attorned to him. On this pretext John claimed that the transaction was invalid, and entered Bainden himself. Without doubt he had overreached himself. He had already been granted Glynde and Patching;[50] in claiming Bainden as well he was asking for too much. Sir William took the side of his brother Richard and initiated an action of trespass against his son in Common Pleas in 1401.[51] The quarrel dragged on for five years until in 1406 the two adversaries submitted to the arbitration of John Ok, the prior of St

[48] PRO, JUST 1/938/1 m. 1.

[49] Edwin Dunkin noted cases in 1392 and 1414 (BL Add. MS 39375, quoting PRO, CP40/528 m. 111d and CP40/615 m. 404d).

[50] GLY 5, 6.

[51] PRO, CP40/562 mm. 106–106d.

Pancras, Lewes, and John Leem, the prior of Michelham.[52] Sir William recovered Glynde and Patching, and his son was compensated with Hawksden, Bainden, and Newenden. They both came out well. The main loser seems to have been brother Richard whose own claim to Bainden was ignored.[53]

Not for the last time, then, a dispute had been solved by what amounted to 'splitting the difference' between the two parties. It was rough justice, certainly, but it was redeemed by an even-handedness that suited this particular kind of case. Far more complex issues were to be submitted to arbitration later in the fifteenth century following the succession of William Waleys III. John III had died in about 1423 before attaining his majority. William, his cousin, was the next male heir, but he was an 'idiot' (as the fifteenth-century pedigree puts it), and his title was contested by John's four sisters. The dispute developed into a classic of its kind, and lasted for the best part of a century.[54] The first attempt at arbitration took place in 1436, when William Chaunterell, sergeant-at-law, and Alexander Aune, recorder of the City of London, found in favour of the four sisters and coheiresses. They were allowed to partition the Waleys estates between them. But ten years later William's claim was again asserted, and an escheator's inquisition *post mortem* named him as the next male heir. After hearings in Common Pleas the case was submitted once more to the judgement of arbitrators. No award is extant, but a later case in Chancery states that 'by mediation of their frendys' it was agreed that William III and his heirs male should have the whole manor of Glynde and that the four sisters and their heirs should have the manors of Hawksden, Bainden, and Patching. By 1460 William had died, and John Waleys IV, a collateral relative who lived in Devon, entered Glynde and sold it to Nicholas Morley, husband of one of the sisters. Morley, if anyone, was to be the victor in this dispute, but his title was challenged in Common Pleas by Robert Lee, husband of another of

[52] GLY 1140 (v). John Leem was, on and off, an important figure locally in the Lancastrian administration. From 1379 until about 1382 he was John of Gaunt's receiver in Sussex, and from 1405 until about 1415 receiver of Pevensey (R. Somerville, *History of the Duchy of Lancaster* (London, 1953), i, pp. 379, 617). The splendid gatehouse at Michelham, which was built during his time as prior, attests far more the greatness of Prior Leem than it does (as is so often said) the vulnerability of the south coast to French raids.

[53] But Richard may have stayed in possession. When John succeeded his father in 1409, he renewed the lease to him (GLY 1235).

[54] The story is told by Dell in his introduction to *The Glynde Place Archives*.

the sisters. The case was transferred to the arbitration of Richard Choke, a justice of the court, and judgement given that Nicholas Morley should retain the manor, but Lee and his heirs take an annual rent of £8. 6*s*. 8*d*. from it. Some years later a further dispute arose about this rent-charge, and this too became the subject of an arbitration award, made in 1498.[55]

In the course of this lengthy and bitterly fought dispute there were no fewer than four attempts at settlement by arbitration, some of them more successful than others. Interestingly, the arbitrators were often lawyers or even judges—a sergeant-at-law and a recorder in 1436, a justice of Common Pleas four decades later. All they were doing was acting in a different (and private) capacity, but doing so, one supposes, at the request of the parties themselves: one of the attractions of arbitration was the chance it gave to the parties to a dispute to name their own umpires.[56] In the dispute at the beginning of the century between Sir William Waleys and his son the two arbitrators, John Ok and John Leem, were presumably chosen one by each side. But the very frequency with which justices and sergeants acted in this capacity, and their willingness to encourage the practice, should remind us that settlement by arbitration was not necessarily perceived by contemporaries (though it might seem so to us) as a substitute for the processes of common law. Quite the contrary, it could be interpreted as the natural outcome of the tendency of those processes to promote compromise. There was a method behind the apparent madness of medieval litigation. The courts were slow, certainly. But it took Maitland's genius to recognize that they had to be, in order to be fair.[57] A minor could not be expected to sue or be sued concerning his inherited rights until he had reached his legal majority; a case could not be expected to go forward in the absence of a jury; and a verdict in an assize could not be allowed to preclude the reopening of the case on different grounds. If the consequence of all these rules was for cases to be drawn out for years, even decades, then that was considered to be a price worth paying. The further that litigants became distanced in time from the events that had produced

[55] GLY 54–8.

[56] There are two good articles on arbitration: I. Rowney, 'Arbitration in Gentry Disputes of the Later Middle Ages', *History*, 67 (1982), 367–76, and E. Powell, 'Arbitration and the Law in England in the Late Middle Ages', *TRHS* 5th Ser. 33 (1983), 49–68.

[57] A point worth stressing, in view of the difficulty of establishing the truth in the middle ages—as the Pashley case shows.

the dispute, the more likely they were to bury the hatchet: to compromise or to submit to the award of an arbitrator.[58] That point was reached in the Pashley case in 1345, when John de Pashley released to his brother his claim to all lands held by the latter in Surrey. There is a nagging suspicion, to be sure, that this 'compromise' was in fact a surrender: that John had little choice but to settle because the scales were so heavily weighted against him.[59] But nearly twenty years after the murders of his father and younger brother there was little else that he could have done.

Compromise, if not necessarily arbitration, was therefore the outcome to which a suit in the courts of common law would most frequently lead. It was also the one most appropriate to the form assumed by most conflict in late-medieval upper-class society. There was a minority of habitual criminals, certainly: and the courts were hopelessly ineffective in coping with them. But most litigants were not criminals, and their wrongdoings were not habitual. Disputes centred on the ownership of land, and were as much intra- as inter-family. Haste was therefore more dangerous than delay, and reconciliation more appropriate than retribution.

Whether by accident or design the courts went some way towards meeting the demands made upon them by fourteenth-century society. Their capacity for change did not admittedly stretch to taking cognizance of the enfeoffment to use. But this seemed not to worry contemporaries: their complaints centred not so much on the structure of the system as on the honesty and integrity of those who practised in it. What the preachers in their pulpits and the Commons in parliament joined together in denouncing was the excessive cost of litigation and the venality of the justices. A petition submitted by the Commons in March 1340 concluded with a request that 'all ministers of each court', including the justices and the barons of the Exchequer, should be sworn to do right to all, without delay and without taking payment from those with business before them.[60] Half-a-century later the peasant rebels were declaring that the land could not be fully free until all the lawyers had been killed; and when that had been done, all

[58] This argument is put forcefully by Palmer, *The Whilton Dispute*, pp. 212–13.

[59] The circumstances are set out in more detail in my article in *History Today*, 34 (Aug. 1984), 30–5.

[60] The petition is printed by G. L. Harriss, *King, Parliament and Public Finance in Medieval England to 1369* (Oxford, 1975), pp. 519–20. See also for an excellent discussion J. R. Maddicott, *Law and Lordship: Royal Justices as Retainers in Thirteenth- and Fourteenth-Century England* (Past and Present Supplement, 4, 1978), p. 43.

things would be regulated by the common people.[61] Their Utopian vision of popular justice remained unrealized. But the reality in the here and how was that more than a few of those same 'common people' were already to be numbered among the thousands of suitors who thronged to King's Bench, Common Pleas, and the assize circuits every year. Doubtless they found the cost of litigation a burden. But whether they found it a crushing burden, whether they regarded a lengthy suit as beyond their means, we cannot yet say.[62]

A study of the sources of our three famlies is hardly going to be a guide to how the poor managed; but it will give an idea of how much the knightly class saw fit to allow for spending on their lawsuits. Appropriately, the fullest information comes from the family which also made fullest use of the courts: the Etchinghams. The early accounts of Beddingham contain under the headings of 'foreign expenses' or 'lord's expenses' a number of payments in connection with litigation that more properly belong to a central account. They record only a proportion of the total spent by the lord on legal fees in any given year; but all the same they are not without interest. Some of the sums represent permanent commitments, or 'overheads' as we might say—for example, the 42*s*. paid in 1310 to Michael de Pyecombe, the lord's attorney in Common Pleas 'pro feodo suo et roba

[61] Maddicott, *Law and Lordship*, p. 62, and the references there cited.

[62] In *Knights and Esquires*, pp. 198–9, where I was arguing that the inadequacies of the courts drove people to take the law into their own hands, I quoted approvingly Professor Sayles's interpretation of a petition which Lewis Cardigan and his wife (seeking the manor of Ladbroke, War.) submitted to the dukes of Surrey and Exeter and four serjeants, in which they sought redress on the grounds that they could not afford the cost of a long action, and that to deny them justice was 'encountre le ley et concience' (*Select Cases in the Court of King's Bench*, vii, ed. G. O. Sayles (Selden Soc. 88, 1971), Appendix VI). But J. B. Post has since shown that the crucial phrase 'encountre la ley et concience' does not meant what Professor Sayles took it to mean ('Courts, councils and arbitrators in the Ladbroke manor dispute, 1382–1400', in *Medieval Legal Records*, ed. Hunnisett and Post, pp. 290, 302). I accept his correction. The problem therefore remains: is there any evidence to show that suitors of moderate means were discouraged by the cost of litigation? I cannot cite any evidence from Sussex; but I think it worth drawing attention to a petition that one James Pabenham, a scion of a Bedfordshire knightly family, but only a 'vadlet' himself, submitted to support his suit for the recovery of the manor of Wilden (Beds.). He implored the king to order the justices of the Bench to render judgement in his favour without delay because he was so impoverished by the pursuit of his right, which he had sought for so long, that he had nothing any longer on which to live or to enable him to continue his suit ('le dit James est tant enpouri par la pursuit de son droit quel il ad si longement . . . qil nad de quoi dount vivere ne la dite pursuite continuer'). Petitions often concluded with a plea of this sort. But since the manor in question was in the hands of James's adversary, his plea here may have been more genuine than most (PRO, KB27/357 m. 20, rotulet 4).

aretro existentibus de anno precedente'.[63] A lord who engaged in litigation as relatively often as Sir William de Etchingham IV did would have found it well worth while to retain an attorney in Common Pleas on a permanent basis. And Pyecombe would have been the obvious man to whom to turn. To judge from his name, he must have been of local origins, perhaps by profession a pleader in the county court, and it is a mark of his standing that he should have been chosen to represent his county in parliament in 1324.[64] Over two years 40*s*. work out as an annuity of 20*s*., which was roughly the going rate.[65] It was a small sum to pay for reliable service and sound advice.

Next there were the expenses incurred in connection with particular suits. These would have varied in amount from one case to another, and no case was typical. It is worthwhile, however, pausing to consider the sums that Sir William disbursed in litigation with another Sussex knight, Sir John de Harengaud, in 1308/9, because these are recorded in some detail in the Beddingham account for that year; and if any set of expenses can lay claim to being considered representative, it is surely this one because it relates to litigation with a man of the same status as Sir William himself.[66] The suit was initiated at a session of the county court held at Lewes, where the jurors were paid 2*s*. 10*d*. to cover their 'expenses', whatever these may have been.[67] Either at that or a later session of the county—its location is not specified—an inquisition was taken in connection with which Sir William made a payment of £1. 1*s*. 9*d*.[68] Subsequently the case must have been transferred to the central courts, because a payment of 4*s*. 1*d*. was now made to one Richard Hervy for seeking writs and 'other necessaries in connection with the said plea'.[69] And two

[63] GLY 996, account roll for 1310–11, 'soluciones forinsece'.

[64] Pyecombe is just north of Brighton. R. C. Palmer, *The County Courts of Medieval England, 1150–1350* (Princeton, 1982), Ch. 4, discusses the careers of men like Michael.

[65] Palmer, *County Courts of Medieval England*, p. 112, says that annuities paid to pleaders in county courts ranged from 6*s*. 8*d*. to 20*s*.

[66] Harengaud held manors at West Dean, Waldron, and Icklesham (*Three Earliest Subsidies*, pp. 3, 21; *VCH Sussex*, ix. 187; BL Add. MS 39374, fo. 149). He represented Sussex in parliament six times between 1306 and 1313, and had died by 1318 (*Cal. Chancery Warrants*, i. 484).

[67] GLY 996, account roll for 1308/9, 'expense nunciorum et forinsece'.

[68] Ibid., 'expense domini'.

[69] Ibid., 'lib' solucio forinsece': Item pagat Ricardo Hervy pro brevibus impetend' et aliis necesariis faciend' pro dicto placito 4s 1d per unam talliam et cedula contra dictum dominum Willelmum Jovene'. Oddly enough, this case has left no trace on the plea rolls of the central courts—or, at least, none that Edwin Dunkin made note of. I have not searched Common Pleas myself.

separate payments of 6*s*. 8*d*. and 18*s*. 8*d*. were made to a couple of attorneys, Roger de Stretton and William de Watergate.[70] In all, the costs incurred in connection with this one suit, and entered on the Beddingham account roll, came to £2. 14*s*. 0*d*.; there may, of course, have been others entered elsewhere or not entered at all.

To a knight as rich as Sir William the sum of £2. 14*s*. 0*d*. was not excessive. But if it had been repeated too often in a single year—and this was probably only a minor suit—it would have made litigation an expensive luxury. Perhaps it is no coincidence, then, that it was the richest of our three families, the Etchinghams, who were the most litigious. They had the widest interests to defend; but they also had the most cash to spend in defending them.

If continuous litigation was not for the poor, neither was it for the uninitiated. By the fourteenth century its complexities had placed it beyond the ken of the ordinary suitor. Professional attorneys assumed responsibility for the procedural moves in litigation, and professional pleaders took over the job of actually speaking for the parties in court.[71] It is the procedural moves that are recorded on the plea rolls, and the pleading that is reported in the Year Books. Assuming that they overlap, these two sources enable us to follow a case in fair detail once it has reached the courtroom. But they tell us little about the trial of strength that was waged between the two parties outside it, little about the pressure that each brought to bear on the other in order to advance his own cause. To savour something of all this behind-the-scenes activity we must turn to the remarkable narrative preserved in the Robertsbridge chronicle of the suit initiated by Simon de Etchingham, the rector of Herstmonceux, for the recovery of the three churches his elder brother had alienated twenty years before.[72]

The churches were those of Salehurst, Udimore, and Mountfield, which William IV had given to the monks in 1314, evidently without his brother's assent, to compensate them for lands of equivalent value they had lost to the sea.[73] According to the surviving fragment of the chronicle (and it is clear that we lack the opening) Simon brought a

[70] Ibid.

[71] Harding, *Law Courts of Medieval England*, p. 78.

[72] The chronicle is BL Add. MS 28550, fos. 2–6ᵛ, printed in translation by C. S. Perceval, 'Remarks on some Charters and other Documents relating to the Abbey of Robertsbridge', *Archaeologia*, xlv (1879), 435–43. Perceval's translation, however, is not always accurate, and I have gone back to the original.

[73] For a discussion of this grant in the context of Sir William's piety see below, pp. 142, 144–5.

writ of *quare impedit* in Common Pleas in the year 1332 to enforce what he regarded as his right to present a parson to the rectory of Salehurst, then apparently vacant.[74] At this stage he was unsuccessful, but he had an influential ally in Robert de Tawton, the keeper of the privy seal, who either had a claim of his own on the living or was expecting to be presented by Simon.[75] Given our ignorance of the earlier part of the story we cannot be certain which. But when Simon brought a second writ of *quare impedit* Tawton was definitely his candidate, because the latter sent letters to the justices of the Bench urging them to exert their influence in his favour. Tawton, however, could not have rated very highly his chances of obtaining the benefice on Simon's presentation, for his next moves were to use his influence with the king to have the abbot summoned by *scire facias* to appear in Chancery to answer the charge that the patents of appropriation had been obtained without royal permission, and to bring writs of *quare impedit* against both him and Simon de Etchingham so as to allow the king to present. How far Edward III was acting on his own initiative or, as the chronicle would have us believe, on Robert's prompting is hard to say; but he did have an interest of his own in the dispute because Salehurst formed a prebend in the royal free chapel in Hastings Castle, to the prebends and deanery of which the crown enjoyed the right of collation.[76] The king and his court were at this time at York, whither the abbot made his way to defend his community's interest, successfully as it turned out, on the technical grounds that the writ required him to answer only on the present king's charter, not that of his father.

While the abbot was fighting these procedural battles, Master Geoffrey de Clare, one of the canons of Hastings, and a future dean, was seeking to arrange a compromise between the monks and the Etchingham family. At a meeting at the manor-house at Etchingham it was proposed that the abbey should pay a stipend of 40*s*. to their vicar in Hastings Castle, and that Simon himself should be bought off by a present of 20 marks sterling and by the provision of a corrody in the

[74] A transcript of the record (Trinity Term 1332) was entered on folio 10^{v} of BL Add. MS 28550. The abbot's defence was that the benefice was in fact full, by the presentation by Sir William de Etchingham of one John de Godele, who was still alive. However, the abbot's attorney was either ill-informed or deliberately disingenuous: John had died by May 1332 (*Cal. Robertsbridge Charters*, no. 332).

[75] For Tawton see T. F. Tout, *Chapters in the Administrative History of Medieval England* (6 vols., Manchester, 1920–33), iv. 77–8.

[76] For the background to this problem see W. R. Jones, 'Patronage and Administration in the King's Free Chapels in Medieval England', *Jnl. of British Studies*, ix (1969), 5.

abbey for William IV's bastard son. All the parties agreed to submit this proposed settlement to the king and to Tawton, but before they had a chance to do so Abbot John was served with a fresh writ of *scire facias* summoning him to York again. He died en route, however, and it fell to his successor John de Wormedale to answer the charges. On his arrival he was asked by the Chief Justice, Sir Geoffrey le Scrope, by what right he had taken possession of a prebend in the king's chapel and how he, not being a canon, could occupy a stall in its choir. The abbot, says the chronicler, feared to reply; but, through the mediation of his friends, an adjournment was secured, and he left for Sussex without judgement being given.

Before he had gone to York the abbot had charged Brother Thomas of Battle to obtain ratification of the draft agreement with the dean and canons and the Etchingham family that had been discussed at the manor-house. This he had been able to do, but only on terms slightly different from those first proposed. Simon's present was increased from 20 marks to 30, presumably on a perception of where this man's true interests were thought to lie, and the provision for the bastard son was altered to a corrody of the victuals of a monk for life, wherever he might be, and without further condition. Simon, in effect was bought off, and it remained to deal with Tawton in similar fashion. He drove a hard bargain. An annuity of 80 marks and no less, he said, was his price for dropping his claim. The abbot, albeit relucantly, agreed; and Tawton then sent instructions to Chancery to halt the action by *scire facias*.

At this point, however, a fresh hitch arose. The Chancellor, John de Stratford, bishop of Winchester, raised once more the problem of reconciling the appropriation of the prebend with the continued performance of services within the king's chapel. He instructed the abbot to attend on him at Farnham Castle; but the date in November that he gave was the very one on which the latter was also due to appear again at York to answer in the adjourned proceedings of *scire facias*. A writ of privy seal was therefore sent to York 'velociter per cursorem' to stay proceedings until Hilary Term, and Master Geoffrey de Clare was dispatched to Farnham with the composition that had been agreed with Simon, the monks and the dean and chapter of Hastings. A few alterations were made, and the Chancellor withdrew his opposition. At last, on 6 December 1333, in terrible weather, Brother Thomas set off for the last time for York with fresh letters under the privy seal to authorize the issue of letters patent confirming Sir William's gift.[77] And in February the following year the abbot, dressed in the habit of a canon, was formally installed in his prebendal pew and assigned a place in the chapter-house in the presence of Simon de Etchingham and other notables.

Here at last we have medieval litigation, warts and all: as it really was, and not as the clerks would have us believe that it was. As a source, of course, the Robertsbridge narrative raises no fewer problems than any other source for legal history. They are very different problems, to be sure, from those encountered in interpreting the plea rolls. The narrative is expansive where the plea rolls are abbreviated, and passionate where they arc detached. In reality, however, these are strengths not weaknesses. Feelings ran high, and here we can feel how high. The abbot and monks had been put to great inconvenience and expense to defend the appropriation of the Salehurst rectory. It was a gift which had cost them dear. Yet in the end justice, or justice of a sort, prevailed. To that extent the cases of *Etchingham* and *Tawton* v. *the abbot of Robertsbridge* can stand for many similar medieval suits that did not find their chronicler.

[77] *CPR 1330–4*, p. 485.

IV

THE ESTATES AND THEIR MANAGEMENT

THAT so much of the violence that characterized late medieval gentry society should have arisen from disputes over the ownership and descent of land is hardly surprising. Land was worth fighting for, worth litigating over. It was, and was for long to remain, the principal source of wealth in pre-industrial society. For some lords the profits of war provided a supplement to income, and for others the rewards of service.[1] But it was rare for anyone with pretensions to gentility to draw the largest part of his income from some other source. Of if he did he soon made sure to invest in land.[2] For the ownership of a country estate brought access not just to wealth but to respectable wealth in the form of lordship over men and acres. And it was the ambition of every landowner to reap therefrom the maximum profit possible.

The ease with which this could be done depended on both the size of the estate as a whole and the geographical distribution of its constituent units. The smaller the estate, the closer the attention the lord himself could give and the greater the chance that it might be run economically. The larger the estate, the greater would be the lord's reliance on professional advisers and the greater, too, would be the opportunities for officials at all levels to line their pockets at his expense. The estates of our three families hardly counted as large by the standards of the greatest proprietors but, in the estimation of their owners at least, they were large and (perhaps more importantly) scattered enough to justify the creation of the same somewhat bureaucratic managerial structure that luxuriated on all the big secular estates of the day. The decision-making process was headed,

[1] Sussex did not rear any great war captains (although the name of Edward Dallingridge of Bodiam is often mentioned in this connection without much supporting evidence). Sir Andrew Sackville and Sir Robert de Etchingham, however, are good examples of local men who grew rich on the rewards of service.

[2] I say this subject to the qualification that there were some professional captains, for whom war had become a way of life, who chose to invest in land only when there was little prospect of military employment. I owe this observation to Philip Morgan.

as always, by the lord himself, acting on the advice of a council composed of the leading officials and perhaps a lawyer or two.[3] Day-to-day management and supervision would have rested with the steward, whose responsibility it was to hold the manorial and (where applicable) hundred courts and to accompany the auditors when they were sent out each Michaelmas to inspect the reeves' accounts.[4] Liveries in kind were sent from each manor directly to the steward of the household and liveries of cash likewise to the lord's receiver. It was from this latter official that the other departments received their allowances. Given the limitations of the evidence there is little we can say about the nature and responsibilities of these several departments beyond speculating that, as in the royal household, they might have corresponded to the lay-out of the offices 'below-stairs' from which they functioned.[5] However, the Etchingham establishment offers one minor curiosity. In the early fourteenth century it seems that Sir William and his wife each had their own receiver.[6] Why this should have been so is not immediately apparent, for in general the establishments of the knights were characterized by greater flexibility, greater informality than those of the magnates.[7] Functional specialization rarely advanced as far in their households as it did in those of their superiors. All the same it remains true to say that, while the titles and responsibilities of officials might vary between employers, the main features of the administrative structure remained the same whatever the size of the household and whatever the extent of the estate.

[3] The Beddingham account for 1366/7 contains a reference to a decision made by Sir William de Etchingham's council (GLY 998/5, 'Liberaciones': 'In liberacione unius porcarii per annum 6 qr 4 bus cap' qr ad 8 septimanas per ordinacionem consilii domini.') For a general discussion see C. Rawcliffe, 'Baronial Councils in the Later Middle Ages', in *Patronage, Pedigree and Power in Later Medieval England*, ed. C. Ross (Gloucester, 1979), in particular, for the role of the gentry, p. 94.

[4] For example, the following expenses are recorded in the Beddingham account for 1376/7: 'Et in expensis senescalli auditorum instauratoris et aliorum pro compotum capiend' de Bedyngham et Pekedean ut patet per unam talliam sigillatam 14s. Et in expensis senescalli pro curiam tenend' per unam talliam sigillatam 1s 4d.' (GLY 1000, Expense senescalli et auditorum').

[5] The Sackville inventory and view of account of *c.*1370 mentions an officer or department of the wardrobe in the family's establishment at Buckhurst. 'Item eadem domina recepit de garderoba de Buchurst per manum Johannis atte Nasshe unum lectum . . .' (SAS/CH 258).

[6] In the Beddingham account of 1310/11 John Bret was described as 'receptor domine' and Richard Agylloun as 'receptor domini' (GLY 996, 'Denarii Liberaciones'). Lady Eve was, of course, an heiress, and it may be that she brought her own servants with her to Etchingham.

[7] As Marie Clough has noticed in 'The Estates of the Pelham Family', p. 192.

That this should have been so is hardly surprising. The knights had served in magnate administrations themselves and knew from personal experience how they worked. In the time of Sir Frank Stenton's 'first century of English feudalism' they had attended the honorial court and advised the tenant-in-chief.[8] In the late-medieval world of 'bastard feudalism' their successors were retained to serve as baronial councillors and were beginning to displace the clerks from many of the administrative offices the latter had long monopolized. Certainly they were displacing them from the stewardship. As we have seen, Sir Andrew Sackville was employed by no fewer than three lords at one time or another in this capacity.[9] Now, if the Andrew Sackvilles of this world were busying themselves managing other people's estates, they could have had little time left to manage their own. To whom, in that case, was this responsibility entrusted?

It is easy enough to compile from the manorial accounts lists of men who served families like the Sackvilles or the Etchinghams in the office of steward or receiver. It is much harder to say anything more about them or to recover any details that might help to bring them to life. Sometimes, just sometimes, we catch a glimpse of their foibles or failings. In 1309, for example, we learn that Simon de Worth, a future steward of the Etchinghams, his brother Henry, and a crowd of others assaulted Robert, the vicar of Burwash, imprisoned him and cut off his testicles.[10] In view of the fact that Sir William de Etchingham was one of the justices of oyer and terminer appointed to investigate the case, it is tempting to wonder if Simon's connection with the family might have gone a good deal further back than can now be traced from the surviving documents. Securing acquittal by the nomination of sympathetic justices was a tactic commonly employed by well-connected malefactors. Simon was not the only client of Sir William's to fall foul of the law. In 1308 Michael de Pyecombe, who was described in the Beddingham account of 1309–10 as Sir William's 'attorney in the Bench', sought a pardon from the king for, as he described it, accidentally killing the daughter of Stephen le Machun of Chailey.[11]

These vignettes add a leavening of picturesque detail; but they hardly take us far in answering the questions we are bound to ask

[8] Sir F. M. Stenton, *The First Century of English Feudalism* (Oxford, 2nd edn., 1961), chaps. ii–iv.

[9] See above, pp. 50–1.

[10] *CPR 1307–13*, p. 246. Simon de Worth is described as the steward in the Beddingham account for 1319/20 (GLY 996, 'Expense senescalli'). He lived at the vill of Brightling, about four miles south-west of Etchingham (*Three Earliest Subsidies*, pp. 207, 319).

[11] *CPR 1307–13*, p. 148.

about the social background of the administrators and the kind of training they received. Pyecombe, as we have seen, was a lawyer who probably mixed the work of an attorney in the central courts with that of a pleader in the county court.[12] But what about the full-time estate officials? Could any of these men have had a legal background? William de Holmestede, Etchingham's steward in the mid-1360s, may have done—he was described as an 'apprenticus de leye' at Cuckfield in the poll tax returns of 1379, but by then he had been gone from Etchingham's service for a good ten years.[13] John Harold who was to succeed him, on the other hand, almost certainly did not. His is an interesting career, and one that is better documented than most. It begins at Etchingham, where we first meet him in the early 1360s in the humble capacity of manorial reeve.[14] From this it may be inferred that he was a local resident and an unfree one at that, because reeves were usually chosen by the customary tenants from their own number. All the same, his humble background proved no hindrance to his advancement. In 1366 he was entrusted with presiding over the manorial court at Udimore, and in the following year with that at Beddingham; and in 1370/1 he is described for the first time as steward, an office he was to hold for the next fifteen years and more.[15] By 1377 he had left Etchingham village for Cliffe, a suburb of Lewes, where he bought a house respectable enough to provide overnight accommodation on at least one occasion for Sir William and his retinue.[16] The

[12] See above, pp. 92–3.

[13] PRO, E179/189/41. Holmestede was present at Etchingham on 9 May 1366 when Sir Robert de Pashley received from his feoffees a quitclaim of the manor of The Moat (*CCR 1364–8*, p. 289). It is interesting to note, therefore, that he was a feoffee of that same Sir Robert, again in 1366 (N. H. MacMichael, 'The Descent of the Manor of Evegate in Smeeth with some account of its lords', *Arch. Cant.* lxxiv (1960), 31). His activities and connections are all very characteristic of the local lawyer.

[14] He is so described in the Beddingham account for 1362/3 (GLY 998/9, 'Porci'). For a discussion of reeves in general, and their opportunities for self-advancement, see below, pp. 125–9.

[15] The Udimore account is in Hastings Museum, JER/Box 3. I am grateful to the curator, Victoria Williams, for allowing the rolls to be temporarily deposited at the East Sussex Record Office at Lewes. Harold is first described as steward in the Beddingham account for 1370/1 (GLY 998/1, 'Expense domini senescalli et auditorum').

[16] The evidence for Sir William staying with Harold is to be found in the Beddingham account for 1377/8: 'Et in probend' equorum domini ad domum Johannis Harold in Clyva 2 bus avene. Et in probend' equorum domini existentis ibidem ut patet per talliam 6 bus.' GLY 1000, 'Avene'). Harold is often now described as 'of Cliffe' (e.g. in the Beddingham accounts and in Lambeth Palace, ED 927, account of the Chamberlain of Ringmer, 1387/8). John Kenting, his predecessor, or predecessor but one, as steward also resided at Lewes, where he paid 4*d*. in the poll tax of 1379 (PRO, E179/189/41). A ludicrous underassessment surely?

demands of his work doubtless made it worth his while to settle in or near the county town;[17] and the rewards that it brought would have provided him with the means—as indeed they did for the other acquisitions of land that he made. These cannot all be traced in detail now, but we know of a meadow and a sheepfold that he held of the archbishop of Canterbury at Ringmer, because he defaulted on the rent that he owed for them.[18] There is just a hint, in fact, that John may not have been too particular about the methods he sometimes employed to advance himself. He and his sons, John and Simon, and two chaplains, John Ingolf and John Hiller, were co-defendants in an assize brought early in Henry IV's reign by Simon Benfield and Cecilia, widow of John Benfield, alleging disseisin of their tenement in South Malling. The sequel is interesting. No judgement was given, because the jury was found to have been arrayed by one Richard Ranger, who was of the counsel of John Ingolf and other of the defendants.[19] Which others? Had Harold himself been involved in rigging the jury? It would not be a great surprise to find that he had: as a past steward he would surely have known how to. Indeed, it is clear from the frequency with which his name crops up that no one was ever more busily employed on his master's business than John Harold had been on Sir William's. When the latter died intestate in 1389, it was therefore appropriate, indeed natural, that he should have been appointed one of the administrators of his goods and chattels.[20] The others were in their different ways knowledgeable men—John Etchingham, Richard Vince, another resident of Cliffe and an Etchingham employee perhaps, and Guy Mone, the archbishop of Canterbury's steward.[21] But it was John Harold who would have known best how to sort things out.

If we return now to the question we asked about the qualifications of the men who administered the knightly estates, the implication, so far as Harold's career is concerned, must be that they had none, or none that involved a formal training—they taught themselves, and learned

[17] He may have moved to Lewes simply for easier access to meetings of the county court. However, it is equally likely (perhaps more likely) that he had been taken on as steward by another employer: the archbishop of Canterbury, perhaps (as lord of South Malling), or the prior of Lewes.

[18] Lambeth Palace, ED 927, account of the chamberlain of Ringmer, 1387/8. It has to be said that no transactions are recorded in the feet of fines. For some reason he chose not to take advantage of the security afforded by this form of conveyance.

[19] PRO, JUST 1/152 m. 54d.

[20] PRO, CP40/517 m. 156.

[21] For Vince, who had been murdered by 1393, see *CPR 1377–81*, p. 546, and *CCR 1392–6*, p. 245. For Mone see Du Boulay, *The Lordship of Canterbury*, p. 394.

from others, as they went along.[22] Perhaps it was these limitations of background and technical know-how that prevented them from climbing higher than they did in the county office-holding hierarchy. John Harold was twice named a commissioner to collect parliamentary subsidies.[23] But that is all. The shrievalty, the bench, even the escheatorship—these richer plums eluded him. Ducal retainers, like the successive generations of the Hungerford family who served the house of Lancaster, could aspire to higher things: but not the servants of knightly families. They were destined to remain middling fish in the small pond of county politics.

Another estate official who found himself drawn into the lower ranks of county office-holding was a clerk by the name of Thomas de Preston, who busied himself with the affairs of a number of knightly employers in the middle years of the century.[24] He started off, it seems, in the service of Sir James de Etchingham, who thought well enough of him to appoint him one of his executors. He remained with the Etchinghams for a year or two after James's death in 1349, but it was Sir Thomas Hoo of Wartling and Sir Andrew Sackville of Buckhurst who were to make the greatest use of his talents in the years ahead.[25] The latter named him as one of his feoffees, always a sure mark of trust, and in 1370 he is found receiving liveries of cash from the estates of the deceased knight at his house at Eastbourne.[26] Bearing in mind the range of connections he had made over the years, it is easy to understand how he came to be appointed a commissioner of walls and ditches in Sussex in 1359 and 1362:[27] and residing as he did at Eastbourne he must surely have known the problems of coastal drainage inside out. But, like Harold, he was never picked as sheriff or escheator. Unlike Harold, of course, he was a man of the cloth, as was his colleague John Spicer, Sackville's receiver and another busy seigneurial official.[28] These two demonstrate by their respective careers that there were still opportunities for clerks as well as for laymen in the world of fourteenth-century administration. Indeed, it is

[22] It is inconceivable that John Harold could have had a legal training, given the strong probability that he was of unfree blood.

[23] *CFR 1383–91*, pp. 69, 116. These commissions were usually composed for the most part of men, like Harold, below the rank of knight or rich esquire.

[24] He is styled a clerk in *Feet of Fines for the County of Sussex, From 1 Edward II to 24 Henry VII*, ed. L. F. Salzman (Sussex Rec. Soc., xxiii, 1916), no. 1959.

[25] PRO, CP40/366 m. 3d; *CPR 1364–7*, p. 69. Wartling account roll, 1359–60 (ESRO, unclassified).

[26] SAS/CH 258.

[27] *CPR 1358–61*, p. 277; *1361–4*, p. 289.

[28] SAS/CH 255, 258.

tempting to wonder if by this time there might have been more openings for them on the estates of the knights than on those of the magnates. Since the latter had been turning to laymen, the clerks may have found that the knights by default had become their main employers. But it is doubtful if the point could ever be proved.

If we are dealing here with estate officials whose opportunities for advancement were limited and whose experience of the world was confined largely to their own county, we are dealing too with men whose work required them to have the same skills as those like Hungerford and Throckmorton who looked after much bigger estates. A steward, whether dealing with five manors or fifty, would still need to be literate enough to cope with the variety of writs that came his way and numerate enough to understand tallies and account rolls.[29] A good steward, like a good bailiff, had to see everything and overlook nothing.[30]

In the age of 'high farming' there was indeed much to be seen and to be overseen. Now that manorial reeves were held to account for every penny that passed through their fingers, an efficient system of audit and account had to be created to replace the kind of informal control that the steward and the receiver had once exercised in person. It could still have been done orally; but in fact it was not. It led to the use of written account rolls, and it is these which form the main source for the present study. They were written by professional clerks hired for the purpose at the rate of 6*s*. 8*d*. a time.[31] Sadly these men never

[29] M. T. Clanchy, *From Memory to Written Record* (London, 1979), interprets the growth of literacy—functional literacy, as Clanchy calls it—as a consequence of the demands of administrative dealings in everyday life.

[30] This was Walter of Henley's advice (*Walter of Henley and other Treatises on Estate Management and Accounting*, ed. D. Oschinsky (Oxford, 1971), pp. 316–17). It would be easy enough to rehearse once more the qualities that Walter and the other writers expected of an ideal steward, but my own preference is to quote the delightful epitaph on the monument to Gabriel Russell (d. 1663) in Tormarton Church, Gloucestershire:

> Here Gabriel Russell lies whose watchfull eyes
> Were William Marquess of Newcastle's spies
> Over three parishes his onely hands
> Were there entrusted with his lordships lands
> Full ninety years my father and I
> Were servants to that Nobillyty
> But all that knew then did them witness bare
> Of their just dealings loyalty and care
> And for their comfort here below
> One and twenty children they could show.

[31] 6*s*. 8*d*. seems to have been the going rate. At Chalvington in 1379/80 however, this sum was struck out at the audit, and 3*s*. 4*d*. substituted (SAS/CH 260).

revealed themselves by name. The question of their identity can therefore only be tackled indirectly by making a comparative study of the hands found in contemporary account rolls from the same estate. If they are all the same, the authorship of a single scribe, hired by the lord, may be assumed. If they are different, then local authorship may likewise be assumed. On the evidence of the account rolls of Udimore and Beddingham, from the Etchingham estate, the only one in our corner of Sussex for which such a study can be made, the latter hypothesis seems the more likely: the two sets of accounts each employ a number of different hands, implying that a succession of local clerks was hired.[32] It is worthwhile as well to compare the hands found in the Beddingham accounts with those in contemporary accounts of the two adjacent manors of Glynde and Chalvington. If, as we have surmised, the Beddingham clerk was a local man, a resident of Lewes perhaps, then his hand might be recognizable in the accounts of other local manors. In fact, however, it is not. The Glynde and Chalvington accounts employ as many different hands as those of Beddingham itself. Evidently there was a pool of talent that could be drawn upon. Should one clerk ever be indisposed or employed elsewhere, there was always another who could take his place.[33]

The fact that manorial accounts were compiled according to a relatively uniform layout by professionals or semi-professionals imposes certain limitations on the use that historians can make of them in reconstructing the pattern of economic change in a medieval village. The clerk's tendency to impose his own logic on the confused and ill-ordered information given to him by the reeve, and to arrange the sources of income and expenditure under a regular set of stereotyped headings has the effect of making manors seem more alike on parchment than they probably were in reality. Admittedly, variations in sources of income could not easily be hidden. But the bureaucratic instinct to make the information conform to the Procrustean bed of accounting convention did its level best to conceal local topographical diversity and differences in field systems. Perhaps this tendency towards uniformity and thus simplification received some encouragement from the use that owners made of the accounts in calculating

[32] My conclusions therefore agree with those of Professor P. D. A. Harvey, *Manorial Records of Cuxham, Oxfordshire, circa 1200–1359* (Oxfordshire Record Society, 1, 1974), 35–42. Professor Harvey was able to identify by name some of the clerks who wrote the Cuxham accounts.

[33] Ibid. 20–1, 36–40.

manorial profits.[34] Dr Stone has demonstrated just how complex and convoluted these calculations had to be, because the accounts were drawn up not, or at least not initially, to establish profit or loss but to show the liability of the accounting official to the lord.[35] Before they could be put to a different purpose some important mathematical adjustments had to be made. And mental adjustments too, no doubt. It is in fact very difficult to capture the view that a secular lord like a knight had of his estates.[36] Did he think of managing them as a whole or as an aggregation of individual manors? The probability is that matters like cropping and payment of wages were left to the discretion of the local reeve. On the Etchingham estates at any rate the sums allowed for the stipends of demesne 'famuli' varied from manor to manor, indicating that levels of remuneration were fixed not centrally for the whole estate but locally by the reeve in response to the conditions of the labour market.[37] The one area of husbandry that does seem to have been run on an inter-manorial basis was sheep farming.[38] Our accounts provide plenty of evidence of flock movements, in these instances over short distances of 10–15 miles or less, but on other estates over much larger ones.[39] Sheep might be driven from one manor to another as they moved up through the stages of adult life. On the Etchingham estates, for example, lambs born at Beddingham were reared at Peakdean, a few miles away on the other side of the Downs, and then driven back to Beddingham twelve months later.[40] Such specialization made sense in an area like the Sussex Downs where sheep-farming was a relatively large-scale enterprise. But thinking in inter-manorial terms rarely manifested itself in other aspects of estate management. Perhaps this was because even on estates of this size manors were usually too scattered for it to offer any advantages in practice. If they lay close together it was a different matter. In the one instance we can cite of two adjacent

[34] Ibid. 15–16.

[35] E. Stone, 'Profit-and-loss Accountancy at Norwich Cathedral Priory' *TRHS* 5th Ser. 12 (1962).

[36] This is what Barbara Harvey seeks to do, and achieves with great success, in respect of a monastic estate, in her *Westminster Abbey and its Estates in the Middle Ages* (Oxford, 1977). As Miss Harvey says on p. v, 'This book . . . is an attempt to capture the view of their estates that the monks had from Westminster over some five hundred years, and to weigh the consequences for their tenants of the thoughts provoked by it.'

[37] For the problems raised by these stipend payments see below, pp. 123–9.

[38] See below, pp. 132–6.

[39] R. Trow-Smith, *A History of British Livestock Husbandry to 1700* (London, 1957), p. 110.

[40] GLY 996.

manors coming into the hands of the same family, they were indeed managed as a single unit. These were the manors of Chalvington and Claverham, the former a long-standing possession of the Sackville family, the latter an acquisition by inheritance of the 1330s.[41] By coincidence the two village economies complemented each other, Chalvington's being predominantly arable and Claverham's predominantly pastoral. After 1337 separate account rolls were still drawn up, but expenses relating to the demesne arable in both villages were charged to the Chalvington account. The Claverham document was left as little more than a stock account.[42]

So much for the accounts themselves. They have their limitations, of course, but we can hardly complain. For lack of anything like them historians of the gentry have been reduced in the past to toying with sources like inquisitions *post mortem* and tax assessments which are too unreliable to be of any value.[43] The result, as Professor Postan once said about the debate on the magnates' fortunes in the fifteenth century, is that verdicts have rested 'not on direct measurement of wealth and income, but on deductions from what we know of the general economic and social situation'.[44] Direct measurement is what will be attempted here. But if our enquiry is not to succumb to the opposite charge of mere parochialism, it must be set within a context provided by Professor Postan's 'general economic and social situation'. A word or two of background at this point may not therefore come amiss.

The situation in the late fourteenth century was by general consent not one that favoured the landlords. Gone were the heady days when land was dear and labour cheap. The Black Death and subsequent visitations of the plague had seen to that. Tenants were harder to find now, and labour more expensive to hire. But it was not the end of the world. For at least a couple of decades after 1348 prices remained high enough to allow labour to be bought on the same terms as before.[45] It

[41] Phillips, *History of the Sackville Family*, p. 85.

[42] For further discussion, see below, pp. 133–6.

[43] Such were the main sources used by the present writer in a chapter on landed income in *Knights and Esquires*. The principal, and very distinguished, exception to my observation is R. H. Britnell, 'Production for the Market on a Small Fourteenth-Century Estate', *Ec.HR* 2nd Ser. xix (1966), 380–7.

[44] M. M. Postan, *Medieval Economy and Society* (London, 1972), p. 175.

[45] A. R. Bridbury, 'The Black Death', *Ec.HR*, 2nd Ser. xxiv (1973), 577–92. For the movement of prices in the second half of the fourteenth century see J. E. Thorold

was only when they came tumbling down at the end of the 1370s that the full effect of the fall in population at long last came to be felt. Profit margins disappeared. Demesnes became a liability. And with every decade that passed more and more landowners were tempted to opt for 'the administratively easier and cheaper policy of leasing'.[46] Manors may have been worth less, and overall income may have stabilized at a lower level; but lords were not reduced to penury. For, if McFarlane was right, they had no need to live of their own: 'wealth attracted wealth; land married land'.[47]

As an outline sketch, that picture might well command acceptance. But it fails to distinguish between various scales of proprietor within the landlord class. Postan made the hypothesis that 'different groups were bound to differ in their ability to benefit and in their liability to suffer from changes in the economic situation'.[48] In the thirteenth century the smaller lay estates, he suggested, possessing as they did smaller demesnes, would have been less well placed than the larger to reap the benefits of the buoyant market for agricultural produce. 'The economic situation should therefore have favoured the magnates rather than the smaller owners'.[49] Arguing on the same lines he formulated a working hypothesis for the subsequent phase, after the Black Death, when wages rose and land values fell. He presumed 'that the class whose income from land took the form of rents or farms must have suffered from the new dispensation; indeed, must have been its main casualty. By the same token the smaller landowners may have suffered less, since most of them consumed a large proportion of their produce and presumably farmed out little, if any, of their land.[50]

Unfortunately, it is less than clear which 'classes' he is talking about. By the 'smaller landowners' he must surely mean those yeoman farmers who by working the land themselves escaped the consequences of the rise in labour costs. They indeed might be said to have suffered least. As Thorold Rogers pointed out long ago, they

Rogers, *A History of Agriculture and Prices in England* (7 vols., Oxford, 1866–1902), ii. 137–67.

[46] J. L. Bolton, *The Medieval English Economy, 1150–1500* (London, 1980), p. 208.

[47] McFarlane, *Nobility of Later Medieval England*, p. 59.

[48] Postan, *Medieval Economy and Society*, p. 160.

[49] Ibid.

[50] Ibid. 174, reproduced in this case word for word from *The Cambridge Economic History of Europe: I, The Agrarian Life of the Middle Ages*, ed. M. M. Postan (2nd edn., Cambridge, 1966), p. 596.

were the principal beneficiaries of the new order.[51] But in that case what happened to the knights? They seem to have slipped through the net of Professor Postan's analysis of the post-Black Death economy. Are they to be numbered among those who suffered badly because their incomes from land took the form of rents and farms? Presumably they are, because their very dependence on rents and farms in the thirteenth century was said to have limited their ability to benefit from the high prices that favoured producers at that time. But if their dependence on rents limited their ability to prosper then, might it not have limited their liability to suffer in the ensuing age of depressed prices? To some extent this distinction between income from rents and income from sales is an artificial one, because the fall in population led to a decline in the demand for land, and thus to a drop in rents, as well as to a fall in prices. But even allowing that rents and prices might have fallen in line, the costs involved in collecting the former were nevertheless far lower and far more stable than those involved in producing demesne crops for the market. And that surely is why, as Mr Bolton has said, landowners in general were tempted to opt for 'the administratively easier and cheaper policy of leasing'. How long a particular demesne remained in hand must therefore have depended on its lord's success in controlling costs, that is to say principally labour costs. Indeed, it has been argued elsewhere that the knights went over to leasing earlier than did the great landowners because the decay of labour services on their estates left them especially vulnerable to the pressure for higher wages.[52] In the light of this hypothesis it is all the more remarkable to find, then, that without exception the demesnes of all the Sussex manors we shall be examining were still directly cultivated right up to the time when the accounts cease—in the case of Chalvington in the 1380s, Beddingham in the 1390s, and Heighton in the opening years of the fifteenth century. These knightly proprietors were still operating as agricultural entrepreneurs at least as long as, and sometimes longer than, the dukes and the earls, the bishops and the monks. On the estates of the earldom of Cornwall, for example, the trend towards leasing had been initiated before the Black Death.[53] After 1348 it gathered pace everywhere. Merton College leased its demesne at Cuxham (Oxon.) as early

[51] Thorold Rogers, *History of Agriculture and Prices*, ii, pp. 667 ff.
[52] Saul, *Knights and Esquires*, pp. 232–4.
[53] E. Miller and J. Hatcher, *Medieval England: Rural Society and Economic Change, 1086–1348* (London, 1978), p. 181.

as 1359.[54] The earl of Norfolk did the same with his at Forncett (Norfk.) before 1375, and the monks of Christ Church, Canterbury the same with all of theirs quite quickly in the 1390s.[55] The abbot and convent of Westminster by contrast withdrew more gradually in stages during the second half of the century, as did the archbishops of Canterbury between the 1380s and 1440s.[56] Moving nearer to our own part of the world, the de Veres leased their small demesne at Laughton in the 1370s, and the monks of Battle put twenty of their twenty-eight manors out to farm—that is, both tenant and demesne land—in the time of Abbot Offington between 1365 and 1385.[57] Yet the knights were still content to struggle on. For our east Sussex families were by no means unique. Over in the west of the county the de Braose family were still cultivating their demesnes in the early fifteenth century.[58] At Acton (Suffolk) Sir Andrew de Bures was cultivating his in the 1390s. And, most remarkably, so too was John Hopton, on and off, on his estates in the same county in the middle years of the fifteenth century.[59]

Any attempt to explain why demesne agriculture survived so much longer than might be supposed on the estates of the knights must depend on an assessment of the function it served in the years before the Black Death transformed the rural scene. Was it undertaken principally with a view to supplying the household with grain? Was the aim self-sufficiency? Or was it geared to production for the market? And what kind of crops were grown? These are difficult questions to answer in general terms because the nature and scale of demesne husbandry varied from manor to manor. Glynde, for example, is the one manor we shall be studying which was a home farm. Here the lord's household absorbed a larger proportion of the harvest than was common elsewhere. At Chalvington, a Sackville manor, no liveries of grain were sent to the household at all in any of the years in the thirteenth century for which we have accounts and in very few of those

[54] P. D. A. Harvey, *A Medieval Oxfordshire Village, Cuxham, 1240–1400* (Oxford, 1965), p. 11.

[55] F. G. Davenport, *The Economic Development of a Norfolk Manor, 1086–1565* (London, 1906), p. 51; R. A. L. Smith, *Canterbury Cathedral Priory* (Cambridge, 1943), p. 192.

[56] Harvey, *Westminster Abbey and its Estates*, pp. 148–51; Du Boulay, *Lordship of Canterbury*, p. 220.

[57] Clough, 'The Estates of the Pelham Family', pp. 72–3; Searle, *Lordship and Community*, pp. 258–61.

[58] W. Hudson, 'On a Series of Rolls of the Manor of Wiston', *SAC* 53 (1910), 144; *VCH Sussex*, vi. i, 264.

[59] PRO, SC6/989/6; C. Richmond, *John Hopton*, pp. 34–5, 79, 84.

in the fourteenth. At Beddingham practice varied from year to year. In 1307 93 qr. of wheat were sold and only one sent to the household at Etchingham, in 1308/9 52 qr. were sold and 62 qr. sent to the household, and in 1309/10 56 qr. were sold and 73 qr. sent to the household. The accounts for the next two years are badly affected by damp, and then there is a break; but when the series resumes in the 1360s we find the same year by year variation. In 1361/2 72 qr. of wheat were sold and 10 qr. sent to the household, and in 1362/3 60 qr. were sold and none at all sent to the household. In 1363/4 by contrast none was sold and 59 qr. were sent to the household. For the following three years the bulk of it was again sold, and then in 1367/8 the proportions were once more reversed. Until the 1370s—by which time, as we shall see, policy had changed—it is fair to say that the wheat harvest was sold in more years than it was sent to the household. But why that should have been so, and on what grounds the decisions were made, we cannot say. For one thing the accounts do not record when during the year the sales were made. Did they wait until they knew that the needs of the household at Etchingham had been met in full? Or were they influenced principally by the level of prices? We simply do not know. The only conclusion we are entitled to draw from the evidence is that until the early-1370s Beddingham was regarded in most years as a cash manor with wheat as its main crop.

The primacy of corn-growing in an area commonly regarded as given over to sheep-rearing was a characteristic that R. A. Pelham noted half a century ago.[60] Even on manors which took in large expanses of Downland it was cereal-growing that formed the bedrock of the local economy. At Beddingham, Glynde, and Heighton, all within a few miles of each other at the foot of the Downs, the main crops were wheat and barley, and at Chalvington, a little to the east on the clay formations, wheat and oats. At Udimore, much further to the east, it was solely oats that was grown in any quantity. This village lay athwart a low Wealden ridge separating the Brede and Tillingham rivers and, as Pelham again observed, cereal-growing in Sussex must be divided into two types, wheat in the chalk zone and oats in the Weald.[61] But—to repeat the important point—it was cereals that predominated on all these manors, both before and after the Black Death.

Of the accounts which have come down to us the most rewarding to study, because the richest in detail, are the long series from

[60] R. A. Pelham, 'Studies in the Historical Geography of Medieval Sussex', *SAC* 72 (1931), 160.

[61] Ibid.

Beddingham. As we have seen, Beddingham was a manor of the Etchingham family and a valuable one too. Its population was not large.[62] But what it lacked in rent-paying tenants—rents of assize only came to £5. 10*s*. 0*d*. per annum before the Black Death—it more than made up for in the size of its demesne, which amounted to some 230 cultivated acres.[63] Wheat and barley, as we have seen, were the two main crops. Wheat was grown for cash, barley for malting and for distribution in wages to the famuli. In the four or five years in the early-fourteenth century for which we have accounts profits from the sale of corn, principally of wheat, ranged from £16 to £40, depending on how much was sold—but anyway a sizeable proportion of a net income usually totalling some £50 or £60. Out of these receipts the reeve had to pay all the routine expenses like weeding, mowing, threshing, and clearing of ditches and in addition to meet a number of charges grouped undifferentially under the headings of 'lord's expenses' and 'foreign expenses' which included retaining fees to local lawyers, payments to attorneys, costs incurred in overnight stays in the village and so on. All told these charges could come to as much as £15 or more each year. Now the timing of their payment is unlikely to have coincided neatly with the collection of rents or the receipt of cash from sales. The reeve would have needed to have a balance on which he could draw. And it is this which is surely represented by the sometimes quite considerable sums styled 'arrears' in the accounts—£3. 9*s*. 0*d*. in 1308/9, £4. 15*s*. 0*d*. in 1309/10, £74 in 1310/11, and £19. 12*s*. 0*d*. in 1332/3. These must represent cash-in-hand.[64]

In the 1370s, however, practice began to change. No longer did the reeve send the bulk of the harvest either to the local market or to the household at Etchingham. A glance at the grain inventory on the dorse of the accounts tells a very different story. It shows that from 1377 onwards a large and increasing proportion of each year's harvest was going instead in assignments to named individuals. There can be no doubt that these *were* assignments and not sales. Sales were always

[62] This statement is based on the fact that there were only 24 taxpayers there in 1327 (*Three Earliest Subsidies*, pp. 200–1). At Udimore, by contrast, there were 57 (ibid. 211).

[63] The total acreage, of course, would have been larger, by perhaps a third or more. We have to allow for land left fallow. The inquisition taken after the death of Sir William de Etchingham in 1326 (PRO, C134/100/15) records 150 acres of arable, 22 acres of meadow and a further 60 acres of parcels of land 'ex perquisito antecessorum predicti Willelmi'. But, as we have already noticed, inquisitions are notoriously unreliable.

[64] The figures are taken from GLY 996 and 996B.

designated as such in the grain account, and the quantity recorded can of course be checked against the appropriate item in the receipts section. So when the latter is seen to decline year on year until it amounts to no more than a few shillings by the late 1380s, there can be no doubt that the harvest was in fact being given away. But to whom and why? The beneficiaries are at least known to us by name. Let the entries in the wheat account for 1377/8 serve as examples:

In seed on 34 acres next to the Furlong 14 qr. In seed on 41 acres of land in diverse plots 15 qr 4 bus. To the sower 1 bus. In seed on 8 acres of land at the Halle 3 qr 5 bus. In reward for the sower for sowing the said 8 acres by order of John Harold 1 bus. Delivered to John Possingworth by order of the lord 21 qr by warrant of the lord. Delivered to Dom Richard Pirington by warrant of the lord 3 qr. Delivered to John Doregg carpenter 2 qr. To Thomas Doregg 1 qr. To Gilbert Carde 2 qr. To John Eggel 2 bus. To John Haycok 1 qr. To John Jolycrop 1 qr. To Robert Sounde 2 qr 5 bus. To Thomas Jolycors 4 bus. To John Halle by warrant 8 qr. To William Tenthe 4 qr 2 bus by warrant. To John Hokman 1 qr. To William Cade 2 qr. To John Harold 2 qr. To the vicar of Beddingham 3 qr. To John Salman 1 qr. To Simon Laurence 1 bus. To Thomas Philip 1 bus. To the Hospital of Seaford by warrant 6 qr. To John Tyghler of Salehurst by warrant 2 qr. To the reeve per annum 6 qr 4 bus. To John Philip reeve of Etchingham 11 qr 4 bus. To Walter Browere 5 qr 4 bus currall. To John Carter 'de mutuo' of the preceding year 2 qr. In expenses of 20 customary works carried out in mowing Broadmead and Southmead 4 bus. In expenses of 3 autumn boon-works 7 bus. Sold as within 15 qr. In advantage for the same, 5½ bus. Sales on account 4 qr 1 bus. And there remain in sheaves 2 qr.[65]

By themselves the names afford but little guide to either the occupation of the recipient or the nature of the transaction in which he had been involved. John Doregg we are told was a carpenter. And John Haycok and John Jolycrop, we know, were minstrels.[66] The 1 qr. of wheat and barley which they each received must have been but a small recompense for the hours of entertainment they had given in the evenings. Other recipients can be identified as administrators—John Harold, for example, and the two household officials Robert Clere and John Possingworth—who were being rewarded indirectly for their

[65] GLY 1000. For an explanation of advantage see R. H. Britnell, '*Advantagium Mercatoris*: A Custom in Medieval English Trade', *Nottingham Medieval Studies*, xxiv (1980), 37–50.

[66] They are described as minstrels in the account for 1390/1 (GLY 998/1). Payments to them began two years previously. They were also given liveries of 2 bu. each of wheat and barley by Sir John Waleys of Glynde in 1372/3 (GLY 1076).

services.[67] Others again can be identified as annuitants, like Gilbert Cooc, the hayward of Udimore in the 1360s, who was given a life-grant of 8 qr. of barley to be taken each year from Beddingham.[68] But this still leaves a large number of recipients who had no traceable connection with the Etchingham administration at all. All we can say is that they seem to have been local men: like Robert Offington, who was the bailiff at Glynde,[69] John Barnaby, who appears in the poll tax returns as a resident of Lewes,[70] and William Cade (a good Sussex name if ever there was one), who was a tenant at Bainden, in the Weald.[71] The explanation for these otherwise mysterious liveries of corn is probably to be found in the last account of the series, that for 1390/1, where we are told that 13½ qr. were given to Thomas Thighler 'pro expensis hospicii', 3½ qr. to one Gilbert 'pro expensis hospicii', and 2 qr. to Thomas Terard and John Toukshe, again 'pro expensis hospicii'. Sir William was choosing to settle his bills not in cash, as his forebears had, but in kind. It follows, then, that his reeve at Beddingham would have had no need any longer to carry the large cash balances he had once held under the guise of arrears for meeting the various payments charged on the manor; and significantly enough these are found to dwindle to only £2 or £3 a year in the second half of the century.[72] Whether or not we choose to call such an arrangement a return to barter, it certainly represents a retreat from the cash economy.

Why this change should have occurred is difficult to say. One possibility is that there could have been a shortage of coin in the 1370s and 1380s. On the other hand, there is no numismatic evidence to support the hypothesis. Indeed, it might be supposed on *a priori* grounds that there would have been more coin per head in the 1370s

[67] Clere is called steward of the household in 1371–2 (GLY 999, 'Multones'). Possingworth is described as a steward (though whether of the estates or of the household is not stated) in 1379/80 (GLY 1001, 'Frumentum').

[68] He appears in the Udimore accounts intermittently from 1364 (Hastings Museum, JER/Box 3).

[69] The Glynde account for 1371/2 is in his name (GLY 1075). He receives large liveries of wheat and barley from Beddingham in 1376–7 and of wheat alone in 1380–1 (GLy 1000, 1001).

[70] PRO, E179/189/41. He receives 3 qr. of wheat and 35 qrs. of barley in 1385/6 (GLY 1002).

[71] GLY 1117, where he is described as a clerk. He received 10 qrs. of wheat in 1384/5 and 4 qrs. in the following year (GLY 1002).

[72] Except for 1381/2, the year of the Peasants' Revolt, when they rose to £13 (GLY 1001). Presumably the reeve ran into problems collecting money.

than there had been in the populous days before the Black Death. Another possibility is that it had become too expensive to send produce to market. In the immediate aftermath of the Black Death carters were certainly difficult to find; and carting services, we know, were among the labour obligations reimposed on some estates in this period.[73] Yet grazing was cheaper in the late fourteenth century than it had been fifty or a hundred years before, and fodder too. Moreover, it is not as if the nearest markets were a great distance away. Lewes was literally only just up the road, and Seaford a mere four or five miles in the opposite direction. The cost of carrying grain to either of these towns could hardly have been so punitive as to force a major re-appraisal of the way the reeve did business. But the decision, it seems, lay not with the reeve or at least not with the reeve alone. The liveries were usually authorized 'per talliam domini', 'pre perceptum domini', or 'per litteram domini'. There was no lack of central supervision. Indeed, one senses that the change of practice was the result of a decision made at the top.

These changes in the method of disposing of the harvest mightily complicate the task of calculating how much the manor was worth to its lord. How much he and his auditors thought it was worth can sometimes be ascertained from the figures for 'profit' entered at the foot of the account. In each case the figure is a high one—£91. 18*s*. 0*d*. in 1310/11, £77. 6*s*. 0*d*. in 1319/20, £51. 11*s*. 0*d*. in 1375/6, £50. 16*s*. 0*d*. in 1381/2 and £58. 13*s*. 9*d*. in 1385/6—but in the absence of any marginal jottings there is no indication of how it was reached.[74] The most we can say is that it is so far in excess of the sum of net income and cash liveries as to suggest that, however abstruse the calculations that were involved in reaching it, they were at least founded on the principle of adding a valuation of produce received to the figure for cash receipts.[75] Our own method must be founded on that same principle if it is to be anything like comprehensive. But it must also take into account that there are bound to be 'leakages'. As we have seen, household bills were paid by means of assignment on the manor, whether in the form of cash in the early part of the century or of corn later on. It is impossible, therefore, to work out a wholly satisfactory

[73] Britnell, 'Advantagium Mercatoris', 50; Du Boulay, *Lordship of Canterbury*, p. 189.

[74] For how the calculations were done at Norwich see Stone, 'Profit-and-loss Accountancy at Norwich Cathedral Priory'.

[75] This was done on the Merton College manor of Cuxham (Harvey, *A Medieval Oxfordshire Village*, p. 94).

'profit' figure of our own to set beside that which the auditors themselves calculated. The truth is too complex to be reduced to a single formula. It is far better to take the sum of foreign payments and liveries to the household in cash and in kind and to see how these change over the century (Table IV). Strangely enough, the trend that emerges is not so very different from the one indicated by the marginal notes of the auditors: that is to say, a period of stability until the late 1370s followed by a decline. The decline looks sharper than it was, because our table does not show the liveries of corn to outsiders which also represented a form of profit to the extent that they saved expenditure in cash. But even the addition of another 10 or 20 qr. of wheat would not disguise the fact that the manor was worth less to its lord in the 1380s than it had been in the 1310s.

Until the late 1370s income (which, excluding wool,[76] means largely income from sales of corn) had held up remarkably well. The Etchinghams' experience at Beddingham provides further support for the view that the golden age of demesne farming came not before the Black Death but in its aftermath in the 1350s and 1360s.[77] In the latter decade receipts stood as high as they had ever done, and in 1370 for the first and possibly the only time they topped £70. It was only with the fall in prices in the late 1370s that decline set in, and thereafter, so far as can be ascertained from the surviving rolls, the lord could only hope to match earlier levels of profit in years of the very highest prices.[78]

Far more curious than the behaviour of income is, in fact, that of expenses. These show no perceptible sign of movement at all during the whole period under consideration—and that in a century when the labour shortage brought about by the Black Death caused a wages explosion. The bill for wages and stipends varied certainly from year to year depending on the relative proportions paid in cash and in kind. But it showed no sign of a secular upward movement in the second half of the century. Small wonder, then, it might be said, that the Etchinghams retained their demesne in hand for so long: there was no reason to think of doing otherwise when they could cultivate them in the 1380s at the same cost as they had in the 1310s.

[76] Income from wool is treated separately below, pp. 132–6.

[77] Bridbury, 'The Black Death', 584.

[78] In 1390/1, for example, when sales of corn realized £22. 6*s*. 0*d*. (GLY 1003). Income from such sales in the 1380s had never exceeded £8 annually. In 1391 wheat from Beddingham was sold at the quite exceptional prices of 10*s*., 12*s*., and 13*s*. 4*d*. a qr.

Table IV Manor of Beddingham

Year	Net income	Sale of wool	Net expenses	Foreign liveries	Cash liveries to household	Liveries in kind Wheat (Qrs.)	Barley
	£s.d.	£s.d.	£s.d.	£s.d.	£s.d.		
1307/8	61 16 7		21 18 9	32 15 0	4 15 8		
1308/9	49 6 2		21 1 7	16 10 10	7 13 5	62	
1309/10	56 3 4	2 8 3	24 1 0	15 3 10	16 16 10	75	
1310/1	56 3 10		24 19 11	8 2	22 8 4	illegible	
1319/20	48 10 10	11 3 4	23 3 2	3 4 4	21 19 10	61	11
1332/3	53 19 2		14 5 11	4 1 9		37	
1361/2	40 6 4		26 16 0		13 16 0	10	10
1362/3	48 12 1		24 7 0	2 0 0	21 10 0	11	
1363/4	34 10 7		21 13 4	4 13 11		59	9
1364/5	62. 13 7	10 9 0	22 11 5		38 9 0	2	22
1365/6	59 17 0		17 19 11		33 14 0	6	
1366/7	60 10 6	9 0 0	22 10 0			6	
1367/8	34 4 0	10 0 0	20 18 7		21 17 5	48	
1368/9	57 9 11	16 13 4	23 9 2		46 8 0	9	
1369/10	35 2 11		22 3 10		3 0 0	10	18
1370/1	72 9 3		21 5 4	6 1	50 14 9	16	10
1371/2	45 0 7		29 18 3	10 9	10 11 0	15	10
1372/3	34 14 5	9 17 8	23 16 11	1 9 0	9 15 8	27	11
1374/5	26 4 11	6 17 4	23 10 6	1 15 9	11 15 0	55	11
1375/6	39 6 6	7 14 0	29 8 10		17 5 0	44	
1376/7	29 1 5		22 2 11	4 19 4		11	10
1377/8	26 3 0	8 17 4	2 13 4	12 11 7	11	13	
1378/9	37 11 4	11 5 6	25 15 9	6 8	13 19 0	7	43
1379/80	28 7 9	10 13 6	26 18 9	2 17 4	10 13 6	31	
1380/1	39 2 2		23 1 2			8	11
1381/2	26 15 1		21 1 4		3 12 6	34	11
1382/3	30 13 2		26 8 6	2 0 0	1 1 0	6	10
1383/4	34 15 11		22 16 4	4 8 2	3 3 6	27	
1384/5	27 13 10		20 15 5	2 12 6	5 13 6		10
1385/6	31 11 1		25 3 9	2 16 10	2 14 0		
1387/8	16 15 0	5 17 1	15 10 6	1 6 10	6 3 9	37	11
1390/1	36 11 1		21 15 2	7 17 8	7 10 0	19	

Notes

[1] The profits from sale of the wool clip from Beddingham and Peakdean were received in some years by the reeve and in others by the lord's receiver. Rather than conflate these sums with the other receipts, thus giving the impression that income fluctuated more than it did, I have entered them in a column of their own.

[2] In the three earliest accounts—1307/8, 1308/9, 1309/10—the paragraph headed 'lord's expenses' contains payments which are in reality not running expenses of the manor but payments charged to the Beddingham reeve rather than to a central account. I have therefore included them under foreign liveries. From 1319 onwards, when they become much smaller and presumably of a more routine character, I have included them in the column of total expenses.

[3] The account for 1332/3 is for 45 weeks only.

[4] The sum of £7.17*s*.8*d*. under 'foreign liveries' in 1390/1 is largely composed of payment of relief on knights' fees due on the succession of the new lord in 1389.

Unless we are mistaken, the Etchinghams' achievement at Beddingham was a truly remarkable one. On this manor (if on no other) they managed to buck the trend. At Udimore, by contrast, the more usual pattern for the period is to be observed. In the course of the 1360s the annual wage bill doubled from £8. 1*s*. 0*d*. to £17. 13*s*. 4*d*. But closer examination reveals that the reason for this was less the rise in the per capital wage than a doubling in the number of men employed. Udimore was a village peopled mainly by tenants of free condition who had never been heavily burdened by obligatory services. So when Sir William decided to expand demesne production, as he did in the 1360s, he could not find the labour force he needed by reviving claims to works that did not exist.[79] He had to take on more staff. And that, in turn, he could only do by matching the wage-rates offered by other employers in the area. He increased the level of stipend in stages, from 8*s*. in 1361 to 10*s*. ten years later.[80] All the same, it was not so much this increase in rates (a very modest one by the standards of the time) but the increase in the numbers employed that doubled the wage bill at Udimore. So long as prices remained high that increase could be absorbed. But if they were ever to fall, there would be problems. In the late 1370s they did fall; but by then inundation by the sea had put paid to demesne agriculture here anyway.[81]

At Beddingham Sir William's position as lord was stronger than it was at Udimore because a far larger proportion of his tenants were unfree and therefore amenable to discipline. On the other hand, the total population was very small relative to the size of the demesne.[82] Given the difficulty that we know every employer encountered after the Black Death in holding on to labour, it is nothing less than

[79] The only surviving Udimore account to record the sale of works in detail is that of 1380/1. There were fewer than a hundred works available, and these were all sold for the sum of 6*s*. Somewhat strangely, however, twenty years earlier in 1361/2 apparently the same quantity of works had been sold for £1. 0*s*. 2*d*. Whatever the reason for this discrepancy, the main point stands: labour services had never figured very prominently in the demesne economy at Udimore (Hastings Museum, JER/Box 3).

[80] For wages and stipends see further below, pp. 122–5.

[81] In 1380–1 the fields of Sagierisland, Rosisland, Emecoteland, Wydesland, Newmarsh, and la Brekke were all noted as being 'sub mare'. So too was the 'Damme'—a dam across the Brede River which carried along its top a path-way to Winchelsea and enclosed a small dock (at least, I take this to be the meaning of 'flota'). In the past this had been farmed for £6. For the effects of weather on other parts of the Sussex coastline in this period see P. F. Brandon, 'Agriculture and the Effects of Floods and Weather at Barnhorne, Sussex, during the Late Middle Ages', *SAC* 109 (1971), in particular Table 2.

[82] Rents totalled no more than £5. 9*s*. 0*d*.

remarkable that here of all places he should have been able to maintain arable production at the level that he did. The area sown never fell more than a fifth below the pre-1348 acreage, and after the absorption of the lands of a deceased tenant in 1376, it came within 20 acres of it.[83] It is even more remarkable that he should have been able to do so at no extra cost. Clearly the reasons for such an exceptional measure of success merit detailed examination. And where better to start than with the labour force itself?

The three main sources of labour on any demesne were the permanent staff (the famuli), the part-time hired workers and the villein tenants of the manor. Distinguishing the relative importance of each of these groups in tilling the soil is easier said than done. The contribution of the villeins can be measured precisely, because the number of days' work they performed was usually recorded on the dorse of the account roll.[84] But the relative shares of the other two groups are harder to quantify, not least because in reality they overlapped. Some of the demesne staff, like the ploughmen, were full-time, but others were hired as and when their services were needed. On the west Sussex manor of Goring in 1346/7 there were 9 labourers working for the full year, 1 for 42 weeks, 2 for 30 weeks, 2 for 19 weeks, 1 for 12 weeks and 7 for shorter periods.[85] We catch here, if only for a moment, a glimpse of the teeming population of semi- or under-employed that filled pre-Black Death England. But few accounts record periods of employment in as much detail as the Goring one does, and even fewer records the names of these faceless workers. Yet without names we stand little chance of establishing how many jobs each famulus actually did: of establishing whether, as at Alciston, for example, the man who did the ploughing at one time of the year looked after the oxen at another.[86] The only account that does give names, the one for Glynde for 1369/70 is disappointing to the extent that it covers only 27 weeks: we are told only what the men did during that limited 6-month period.[87] But if it fails us in that respect, it helps

[83] Between 1307 and 1311 roughly 210 acres were sown each year. By the 1360s the acreage had fallen to about 170–80, but it went up to 190 after 1375, when John atte Halle's tenement was included.

[84] The word 'usually' is worth stressing. There is, for example, no such record on the dorse of the accounts for Glynde, although it is clear from the paragraph of autumn expenses that boon works and perhaps other services were exacted both before and after the Black Death.

[85] BL Add. Roll 56359. The manor belonged to the Tregoz family.

[86] J. A. Brent (née Wooldridge), 'Alciston Manor in the Late Middle Ages' (Bristol MA thesis, 1965), p. 64.

[87] GLY 1074.

us in another. It sheds some light on the origins and backgrounds of the famuli. Some of them were born and bred in the village. John Boylyng, the master famulus, came from an old and by then widely ramified Glynde family.[88] So too did Henry Pocock, the shepherd. But several of the others hailed not from Glynde but from the Waleys properties near Mayfield. William Aleyn, a ploughman, probably came from Hawksden.[89] William Bainden and William Whebenham certainly came from Bainden itself because their names crop up in the rental of that property compiled in 1363.[90] How these men had come to be employed on the demesne at Glynde is an intriguing question. That labourers should have been on the move in the post-Black Death world is only to be expected. They went to work for whoever would pay them most. But moving from one manor to another owned by the same employer is not the same as changing employers. Could they have been attracted to Glynde by the prospect of being paid higher wages there? We know that wages did vary between manors of the same estate.[91] Could they have been drafted to Glynde? Or had the demesnes, if any, at Bainden and Hawksden been put out to lease? We simply do not know.

In broad terms the number of famuli employed varied with the size of the demesne—the larger it was, the greater the number of staff that would be needed, and vice versa. At one extreme was Goring, with its enormous demesne of nearly 600 acres, worked by nine full-time ploughmen and a host of part-time staff. At the other was West Dean, with no more than 80 acres under seed in 1387 and only two ploughmen employed.[92] But the relationship was rarely as mechanistic as this because of the existence of a third variable, the quantity of labour performed by the customary tenants of the village. Over a period of time the relative proportions of these several sources of labour would have changed constantly. At Glynde, for example, the number of famuli employed fell so much more sharply after the Black Death than did the cultivated acreage of the demesne—the former by a half between 1346 and 1368 and the latter by only a fifth—that one can only suppose that the work was done, if it was done at all, by making

[88] The family name crops up several times in a rental of Glynde compiled in 1398–9 (GLY 1059).
[89] The lease of Hawksden was taken in 1394 by one Geoffrey Aleyn (GLY 1223).
[90] GLY 1117.
[91] For a discussion of wages and stipends, see below, pp. 122–5.
[92] SAS/M/673. The manor belonged to the Poynings family.

increased demands upon the services owed by the customary tenants.[93] We know that such services did survive well into the late fourteenth century; but usually those taken up were not so much the highly unpopular week-works like ploughing as the seasonal boon works and threshing services.[94] Only at Beddingham was ploughing required of the customary tenants, and then not in sufficient quantity to make much impact on a demesne of nearly 200 acres. If, as we may suspect, the amount of labour available—whether permanent, casual or customary—was a good deal less in the 1380s than it had been half a century before, Christopher Dyer may well be right in suggesting that ploughing resources were not always adequate for the task of cultivating the whole of a demesne.[95] The skimping of this and other labour-intensive tasks may well have been one reason for the low yields characteristic of the period.[96]

If this hypothesis is correct, there was little remedial action the lord could take short of employing more people. And that he would have been reluctant to do seeing that profitability depended so much on successful control of wage costs. So how effectively were these costs controlled? The demesne staff were paid partly in cash and partly in

[93] There seem to have been eight full-time ploughmen in 1347–8 and only three twenty years later. However, three more famuli were taken on for the winter of 1368/9—from Michaelmas to 22 April, to be precise—and, although it is not specifically stated, these may well have helped with the ploughing. In 1347/8 the harrower was also said to have helped with the ploughing in the summer and autumn. The area sown declined from 250 acres on the eve of the Black Death to 200–10 acres in the 1370s.

[94] Harvey, *Westminster Abbey and its Estates*, pp. 258–9, for example. Degrees of unpopularity are of course relative. It may have been ploughing services which provoked the fiercest resistance, but other types of compulsory labour were scarcely performed with any enthusiasm. In 1373 there was a presentment in the manorial court at Beddingham 'quod Thomas Jakelot inobediens erat in autumpno metendo bladum domini' (GLY 980).

[95] C. Dyer, *Warwickshire Farming, 1349–c.1520, Preparations for Agricultural Revolution* (Dugdale Soc. Occasional Paper, 27), pp. 14–15. I find this explanation more convincing than Searle's hypothesis of increasing efficiency (*Lordship and Community*, p. 307).

[96] I base this statement on calculations I have made using the Beddingham accounts, the only ones to provide a long enough run of consecutive accounts to justify the exercise. The gross yield of wheat per seed ranged between 3.2 and 5.8 in the period from 1362 to 1386, with 4.5 being about the average. The gross yield of barley ranged over the same period between 1.9 and 4.1 with three times being roughly the average. These figures are comparable with those recorded by Dr Titow from the Winchester estates, 1200–1350 (*Winchester Yields: A Study in Medieval Agricultural Productivity* (Cambridge, 1972), in particular Appendices C and D). If anything, the barley yields at Beddingham were worse than those obtained on the pre-Black Death Winchester estates.

kind: they would receive a yearly stipend or retaining fee and a weekly allowance of corn or a cash wage ('vadium') of corresponding value if the corn was in short supply.[97] Any attempt to calculate the total annual wage bill on a manor is bound to involve estimating the monetary equivalent of this corn livery, an exercise which in its turn involves making assumptions about the price each year of different grades of corn. The calculation can be done, but only at the cost of building one assumption upon another. Perhaps it is better, and certainly just as instructive, to trace changes not in the total wage bill but in the levels of cash reward paid to two important manorial servants, namely the plough-holder and the shepherd. The evidence from the accounts of Glynde, Chalvington, Udimore, and Beddingham is presented in tabular form.

Table V Cash wages of demesne staff at Glynde

	Wages of a ploughman		Wages of a shepherd	
Year	Weekly vadium	Annual stipend	Vadia	Annual stipend
1347/8		7*s*.		4*s*.
1368/9	7*d*.	7*s*.		
1371/2	7*d*.	7*s*.	6*d*.	6*s*.
1372/3	7*d*.	5*s*.*	6*d*.	4*s*. 6*d*.

* Not a full year.

These figures are a far cry from the wage explosion the anguished petitions to parliament might have led us in our innocence to expect. Even at Udimore, the manor on which, as we have seen, the lord's need to match the going rate was greatest, the ploughman's weekly wage remained unchanged in the 1360s and his stipend climbed by no more than 2*s*. At Beddingham in the same period the stipend rose by only 6*d*. to 8*s*. 6*d*., and the figure of 10*s*. allowed at Udimore for the first time in 1364 was not reached until 1390.

The auditors can sometimes be seen at work in the substitution of a lower rate for a higher one. For three successive years at Beddingham in the 1370s a figure of 9*s*. claimed by the reeve was struck down and replaced by the established one of 8*s*. 6*d*. But even this more modest

[97] This is the explanation given in the 'Vadia' paragraph in the Beddingham account for 1384/5 (GLY 1002).

Table VI The cash wages of Demesne Staff at Beddingham

	Plough-holder		Shepherd	
Year	Weekly vadium	Annual stipendium	Weekly vadium	Annual stipendium
1307/8	6*d.*	7*s.* 6*d.*	6*d.*	4*s.*
1308/9	6*d.*	8*s.*	6*d.*	4*s.*
1309/10	6*d.*	7*s.* 4*d.*	6*d.*	4*s.*
1310/11	7*d.*	8*s.*	7*d.*	4*s.*
1319/20		8*s.*		4*s.*
1332/3		8*s.*		
1361/2	7*d.*	8*s.*	7*d.*	4*s.*
1362/3	7*d.*	8*s.*	7*d.*	4*s.*
1363/4	7*d.*	8*s.*	7*d.*	4*s.*
1364/5	7*d.*	8*s.*	7*d.*	4*s.*
1365/6		8*s.*		4*s.*
1366/7		8*s.* 6*d.*		4*s.*
1367/8		8*s.* 6*d.*		4*s.*
1368/9		8*s.* 6*d.*		4*s.*
1369/70		8*s.* 6*d.*		4*s.*
1370/71		8*s.* 6*d.*		4*s.*
1371/2		8*s.* 6*d.*		4*s.*
1372/3	7*d.*	8*s.* 6*d.*	7*d.*	4*s.*
1374/5	7*d.*	8*s.* 6*d.*	7*d.*	4*s.*
1375/6	7*d.*	8*s.* 6*d.**	7*d.*	4*s.*
1376/7	7*d.*	8*s.* 6*d.**	7*d.*	4*s.*
1377/8		8*s.* 6*d.**		4*s.*
1378/9		8*s.* 6*d.*		4*s.*
1379/80†				
1380/81	7*d.*	8*s.* 6*d.*	7*d.*	4*s.*
1381/2	7*d.*	8*s.* 6*d.*	7*d.*	4*s.*
1382/3	7*d.*	8*s.* 6*d.*	7*d.*	4*s.*
1383/4	7*d.*	8*s.* 6*d.*	7*d.*	4*s.*
1384/5	5*d.*‡	8*s.* 6*d.*	5*d.*	4*s.*
1385/6	5*d.*‡	8*s.* 6*d.*	5*d.*	4*s.*
1387/8		8*s.* 6*d.*		4*s.*
1390/91		10*s.*		8*s.*

* The figure of 9*s*. is deleted by the auditors and replaced by 8*s*. 6*d*.
† The account for the year is illegible.
‡ The figure of 7*d*. is crossed out by the auditors and replaced by 5*d*.

rate was higher than that allowed at the neighbouring manors of Glynde and Chalvington, where the plough-holders were only permitted 7*s*. and 6*s*. respectively. The Etchinghams' achievement in this difficult period, therefore, was not unique—the Waleyses and the Sackvilles were no less firm with their own staff—but all the same it

Table VII The Cash Wages of Demesne Staff at Udimore

	Wages of a ploughman (plough-holder)	
Year	Weekly vadium	Annual stipendium
1361/2	7*d.*	9*s.* 6*d.*
1362/3	7*d.*	8*s.*
1363/4	7*d.*	8*s.*
1364/5	7*d.*	10*s.*
1365/6	7*d.*	9*s.*
1366/7	7*d.*	9*s.*
1368/9	7*d.*	10*s.*
1369/70	7*d.*	10*s.*
1370/71	7*d.*	10*s.*
1380/81	7*d.*	6*s.* for ¾ yr.
1381/2	7*d.*	6*s.* for ¾ yr.

Table VIII The Cash Wages of Demesne Staff at Chalvington and Claverham

	Wages of a ploughman (plough-holder)		Wages of a shepherd	
Year	Weekly vadium	Annual stipend	Vadium	Annual stipend
1279/80	6*d.*	6/8*d.*		
1294/5				
1337/8				
1338/9		6*s.*		
1346/7		6*s.*		6*s.*
1349/50	6*d.*	6*s.*	entry deleted	
1364/5		6*s.*		5*s.*
1366/7	7*d.*	6*s.*		
1379/80	7*d.*	8*s.*		

was exceptional. In the 1380s a reasonable rate for a plough-driver, who normally earned less than a plough-holder, was reckoned to be at least 10*s*. and the rate that he could command in the market was commonly far higher. In Essex 26*s*. 8*d*. was the amount most often paid.[98] Nearer home, on the manor of Marley, which belonged to Battle Abbey, the stipend was 10*s*. in the 1360s and 17*s*. in the 1380s.[99]

[98] N. Ritchie, 'Labour Conditions in Essex in the Reign of Richard II', *Ec.HR* iv (1934), reprinted in *Essays in Economic History*, ii (London, 1962), pp. 91–111.
[99] Searle, *Lordship and Community*, pp. 304–9.

At Mayfield rectory the archbishop of Canterbury was allowed to escape with paying a more modest 13*s*. 4*d*.[100]

The disparity between the levels of remuneration granted by these other employers and those recorded in our accounts is so great as to arouse the suspicion that the latter may not correspond to what was actually paid by the reeve to his men. Some evidence of such 'wage drift' is afforded by the deletions made at the audit. But these do not occur often enough to suggest either that pressure from above was stronger on these estates than elsewhere or that it was this which was responsible for the depression of wages. In order to understand what was happening we must turn for a moment from expenses to the works' accounts which were recorded on the dorse of the Beddingham compotus rolls.

At the beginning of the fourteenth century the total number of works nominally available at Beddingham was 1,470. By 1361, when the accounts resume after a long interval, it had come down to 945. The fact that the size of this reduction—roughly one-third—corresponds so closely to the rate of mortality generally assumed to have been caused by the Black Death suggests that the quotas were reassessed after its passing to take account of the fall in the number of tenants occupying land held in villeinage.[101] Yet at the same time the sum recorded in the charge section as due in 'rents of assize' remained unchanged at between £5 and £6, suggesting (improbably) that the number of tenants had not fallen at all. On some estates a shortfall in the amount actually received was shown by entering 'decays of rent' in the discharge section—but not here at Beddingham. There was no attempt to bring the accounts back into line with reality.[102]

But with what reality? It had never been the object of accounting to establish profits or loss as such. Its primary purpose, as Professor Harvey has written, was to determine the state of account between the proprietor and his official: which of them was in debt to the other and by how much.[103] The additions, deletions, and substitutions that occur in every compotus roll, even in 'fair' copies, demonstrate beyond doubt that the process of settling an account was a lengthy business, a

[100] Lambeth Palace, ED 695.

[101] This hypothesis is strengthened by the fact that a new rental was compiled in 1351. This is GLY 991.

[102] It is worth mentioning, too, that the chevage payments look suspicious. There is an item of 4*d*. payable by one John Leyman from 1361, when the accounts resume, to 1375, when it finally drops out.

[103] Harvey, *Cuxham Manorial Records*, p-. 14–15.

battle of wits which involved many hours of difficult negotiation between reeve and auditors. The balance that was finally struck between them was as much a reflection of their relative bargaining strengths as of any economic reality that existed on the ground. It follows, therefore, that the wages and stipends entered in the discharge section were not necessarily the actual sums disbursed by the reeve to his men. They were simply the amounts that the auditors allowed him for that purpose, and whatever he had to pay in excess in order to retain staff he was expected to meet out of his own pocket.

In an age of sharply rising labour costs the reeve could in theory, then, have left office poorer than he entered it; indeed, when we recall too the likelihood that he would not have been able to collect the rent charge in full, it becomes a source of wonder that anyone was willing to serve in the office at all. Evidence of men's reluctance to do so is to be found in the fines that they sometimes paid to secure exemption. In 1333, for example, John Cook and John Hykeman each paid 12*d*., and John le Sounde as much as 6*s*. 8*d*., 'ut absolvatur hac vice de officio prepositi'.[104] Such fines were still paid in the late fourteenth century, but between times their significance had changed. The villagers, instead of nominating a single candidate to serve, as they had in the past, now proposed two, of whom the lord or his steward chose one. The unsuccessful candidate was then amerced, usually in the sum of 2*s*. Here as elsewhere, therefore, the method of choosing the reeve was a mixture of election and selection.[105] But in the 1370s selection took over completely, and the lord began to ignore altogether the formality of the election in favour of appointing someone himself. In July 1369 William Philip and John Parkyn had been nominated, and Parkyn was appointed to serve.[106] But twelve months later, when William atte Leye and Peter Sounde were nominated, neither was appointed and Parkyn was left in office—presumably because he had done a deal with the lord which guaranteed him enjoyment of the office for a certain length of time in return for an underaking to deliver to the household a mutually acceptable annual surplus in cash and in kind. Parkyn kept the office for a period of three years, at the end of which he handed over to Peter atte Welle. Two years after that the villagers' wishes were again overridden. Robert Sounde and Thomas Philip were nomi-

[104] GLY 973.

[105] Christopher Whittick tells me that at Herstmonceux, too, the lord chose one of the two nominees put foward by the villagers in the manorial court.

[106] GLY 977 (court of 28 July 1369).

nated, but Peter atte Welle was re-appointed, and kept in office for the next two years.[107]

If men were prepared to serve (as they were) for periods of two, three, even four, years, the perquisites of office both official and unofficial must clearly have been considerable. At Beddingham the official rewards were reckoned as remission of the rent and services due from the reeve's holding and payment of an emolument of £2. 12*s*. 0*d*. per annum: benefits which compared well with those offered on other manors in the fourteenth century,[108] but which would scarcely have gone far to recompense the reeve for the extent to which he had to top up the labourers' wages out of his own pocket. Simply to break even, let alone to make a profit, he would have had to improve on that. It was as well, therefore, that fortune gave him no shortage of opportunities to do so.

The unofficial, perhaps illicit, rewards of office are by their very nature hidden from view, but it is not difficult to imagine the form that they took. It would be surprising, for example, if the reeve had not folded his sheep on the lord's pastures or used the demesne stock to plough his own lands.[109] In an age when land was falling into disuse, it would be even more surprising if he had not snapped up tenements as they fell vacant or organized exchanges of land so as to assemble a more convenient little 'estate' for himself. And, as Drew pointed out long ago, he would have been better placed than most of his fellow agriculturalists to hear of opportunities for making profitable purchases and sales in connection with his own business.[110] In short, possession of the office gave a tenant the chance to better himself economically and to steal a march on his rivals in the village.[111] It was an indispensable adjunct to a successful career.

[107] GLY 980 (court of 12 Oct. 1374).

[108] On the Winchester estates the reeve was given an allowance of 9*d*. to 12*d*. a week, remission of rent and labour services and food at the lord's expense during the harvest period: concessions which J. S. Drew estimated to have been worth about 24*s*. per annum (J. S. Drew, 'Manorial Accounts of St Swithun's Priory, Winchester', *EHR* lxii (1947), reprinted in *Essays in Economic History*, ii. 28). At Cuxham the reeve did not even get an emolument. He was allowed simply remission of rent and labour services (Harvey, *A Medieval Oxfordshire Village*, p. 70).

[109] As P. D. A. Harvey points out, there seems to have been a good deal of give and take in the reeve's use of his own stock and that of the demesne (*A Medieval Oxfordshire Village*, p. 70). I am grateful to Professor Harvey for advice on a number of points concerning the interpretation of manorial accounts.

[110] Drew, 'Manorial Accounts of St Swithun's Priory, Winchester', p. 28.

[111] It should be remembered that John Harold, Sir William V's long-serving reeve in the 1370s and 1380s, began his career as reeve of Etchingham. See above, pp. 101–2.

The reeve's position, therefore, was by no means as circumscribed as it might appear to have been. The reluctance of the auditors to come to terms with the rise in wages impelled him to be all the more predatory in his day-to-day dealings with his fellow tenants; but the less rigorous auditing to which he was subjected allowed him to be so without fear of censure. No wonder bailiff and herdsman alike were 'adrad of hym as of the deeth'.[112]

The more independent the reeve's position became, the more his relationship with the lord came to resemble that of a lessee. On some estates, indeed, he was required to answer annually not for the entire issues of his manor but for a predetermined yield of cash or of crops—in other words, a rent, or farm, in all but name.[113] But it does not seem that this was so on the estates of the Etchinghams. As Table IV shows, there was no regularity in the liveries paid over by the reeve at Beddingham. Even in the years after 1380, when the cash livery declined to a mere pittance, the corn livery varied from nearly 50 qr. one year to 20 qr. the next.[114] It seems more likely that here it was not the lord but the reeve who had the satisfaction of knowing in advance how much he was going to receive each year. The prominence of the well-do-do in the office[115] and the irregularity of payments to the lord can only be explained on the assumption that it was the reeve who had the fixed return and the lord the surplus. If that sounds an unlikely arrangement, it is nevertheless the only one which accords with the facts as we have them. The days were gone (if they had ever existed) when the reeve answered for all the profits. To that extent the last phase of direct management anticipated the first phase of leasing. The transition from the one to the other was by no means as sharp as might be supposed.

[112] *The Works of Geoffrey Chaucer*, ed. F. N. Robinson (2nd end., Boston, 1957), General Prologue, line 605.

[113] Harvey, *A Medieval Oxfordshire Village*, p. 74.

[114] There was no more stability in liveries from the manor of Udimore in the period covered by the surviving accounts. But that was a consequence of first the expansion and then the contraction of agriculture on the demesne there.

[115] There can be no doubt that the reeves were for the most part recruited throughout the fourteenth century from the four or five most affluent families in the village. The Coocs, atte Welles, atte Leyes and de Soundes were filling the office in the 1370s and 1380s as they had been in the 1310s. In 1333 Thomas atte Welle had contributed 5*s*. to the twentieth granted in parliament, John atte Leye 4*s*., and John Cooc 3*s*. A couple of villagers paid higher sums; but all the rest paid less (*Three Earliest Subsidies*, pp. 200–1).

So far we have been concerned primarily with the affairs of the Etchinghams, and it remains to be seen how far their experience was typical of that of other landowning families in east Sussex. It has to be said that for no other manor is the documentation as rich as it is for Beddingham or Udimore; but all the same a few generalizations can be made.

We need only look at the one surviving account roll from the Poynings estates in Sussex to see that, while seigneurial policy was much the same everywhere, there was local variation in the manner of its application. At West Dean, near Seaford, as on all the other manors we have discussed, the demesne was still in hand in the late 1380s. But here Sir Richard Poynings seems to have kept his reeve on a tighter rein than the Etchinghams did theirs. The charges for rents of £3. 12*s*. 0*d*. and for farms of £1. 2*s*. 6*d*. were balanced in the discharge section by an item for 'allowances and decayed rents' of no less than £2. 4*s*. 0*d*.[116] The size of this item, amounting as it does to nearly half the combined total of rents and farms, suggests that the Poynings' auditors were more insistent than the Etchinghams' on the account rolls being kept up to date.

But if so they were probably the exception. On the Sackville manor of Chalvington there must be a strong presumption that the accounts became as unreal as they did at Beddingham. Stipends were even lower than those allowed there. Rents of assize never varied over the whole century; and 'decayed rents', which made their first appearance in 1349, soon settled down to a suspiciously unvarying 10*s*. 6*d*. The tell-tale signs are all to be seen. But in the absence of any works' accounts internal inconsistency cannot be demonstrated here as conclusively as it could on the Etchinghams' manor.

At Heighton St Clair, on the other hand, it probably can. No more than a handful of accounts survive, but happily they come from opposite ends of our period. In the year 1285/6 572 works were available, of which 453 were taken up. A century and a quarter later 542 were available of which 298 were taken up.[117] It is interesting to see that here, as at Beddingham, works were still being used at so late a date. But it is more to the point to observe that here as well a reduction in the number available had occurred, albeit this time a far smaller one than at Beddingham. Over the same period, meanwhile, the charge for rents of assize had actually crept up slightly from £14. 18*s*. 6*d*. to £16.

[116] SAS/M/673.

[117] SAS/G1/44–6.

It had not been swelled by sales of works: these were entered in a separate paragraph. It had presumably edged up in the early fourteenth century, when population was increasing, and had not been adjusted when it fell. That population did fall, and fall very quickly by far more than the decline in labour services would suggest, is proved by the eventual disappearance of the village. But so little do the accounts correspond with reality that they scarcely betray the existence, let alone the scale, of the demographic catastrophe that was to engulf the settlement a couple of centuries later.[118]

At Glynde, too, the charge for rents of assize remained virtually unchanged into the post-Black Death era, at £12. 4*s*. 3*d*. in 1347/8 and £11. 7*s*. o*d*. in 1368/9,[119] and suspicions are aroused by the absence from any of the later accounts of a section of 'allowances or decayed rents'. But on this manor we can go beyond mere suspicion because the survival of two rentals, one compiled in *c*.1290 and the other in 1353, provides an independent basis of comparison.

The earlier of the two, compiled in the time of Sir Richard Waleys, identifies some 48 tenants in Glynde and Glyndebourne and three more in Millink (in Ringmer), bringing the total to 51.[120] By the time the second was compiled, some five years after the Black Death, the total had fallen to 41.[121] But this relatively small drop in numbers concealed a big change in the composition of the population. Of the four dozen or so names recorded in the rental of 1290, twenty-two had disappeared by the early 1350s—whether as a result of the generally high rate of extinction in the male line in the middle ages or as a direct consequence of the Black Death we cannot be sure, but disappeared they had. Who, then, were their successors? Their names are not ones recogniable from the earlier rental. They could have been people once landless who had acquired land for the first time. Or they could have been newcomers to the village. Some at least came into this latter category—Henry Bainden, surely (to judge from his name), who was the reeve at Glynde in 1347/8, and John Dawe, whose name crops up

[118] Its site is today occupied by a wide expanse of Firle Park (I. D. Margary, 'Roman Roads from Pevensey', *SAC* 80 (1939), 53). It had a church; and with a rental income in the late thirteenth century of £14. 18*s*. o*d*., nearly three times that of neighbouring Beddingham, it was clearly, in its heyday, a not inconsiderable place. There were still 21 taxpayers left in 1524/5 (*Lay Subsidy Rolls for the County of Sussex, 1524–5*, ed. J. Cornwall (SRS 56, 1956), p. 117).

[119] GLY 1072, 1073.

[120] GLY 1058. This figure of 51, it should be remembered, is the number of rent-paying tenants. A multiplier should then be employed to obtain a rough idea of the population of the vill. However, I do not intend to risk my neck by suggesting what the multiplier should be.

[121] GLY 1061.

in the Bainden rental of 1363.[122] Other newcomers can be identified not by place of origin but by occupation—William Bush, Sir John Waleys's cook, and Richard Daleham, his park-keeper.[123] These were all men who had been drawn to Glynde by the nature of their employment and once there had settled down. The tenements they took were ones that had been held in villeinage in 1290 but which in 1353 were described as acquired 'per cartam domini de novo'. They had been re-let, presumably, on terms that would have made them more attractive to possible tenants. Heriot and relief are usually mentioned but not, significantly, labour services.[124] The tactic worked: enough tenants were attracted to take most if not all of the holdings that had fallen vacant.[125] For the moment, then, Sir John could relax; but it would be unwise to conclude that he was delivered from the problems that beset the Etchinghams at Beddingham or the St Clairs at Heighton. The 1350s were not the 1380s, and seigneurial success in the earlier decade was no proof against failure several plague visitations and many mortalities later. Certainly Glynde had been, and was always to remain, a more populous place than either of the other two vills. But the high turnover in ownership of tenancies revealed by comparison of the two rentals should both remind us not to be deceived by stability of rental income and alert us to the dangers that lords were going to face in the 1380s and 1390s when the reservoir of landless dried up.

In some parts of England, notably the east Midlands, landowners faced with the disappearance or virtual disappearance of their tenants

[122] The tenement which Henry de Bainden is recorded as holding in GLY 1061 must be the one formerly held by Ralph le Hone in villeinage which Sir John Waleys granted by indenture to Bainden and his wife Alice on 12 June 1340 (GLY 1163). Likewise, Dawe's tenement must be the one which Sir John granted to him and his wife by indenture on 19 July 1347 (GLY 1165).

[123] Bush appears in GLY 1073. Daleham appears in all the later accounts (GLY 1073–6). Daleham acquired land in a field called Westhurst from Sir John in September 1350 (GLY 1166, 1170) and pasture in Glyndebourne from Richard Bost, another villager in 1357 (GLY 1176). He was dead by 1389 (GLY 1189).

[124] These Glynde leases bear comparison with those made in the same period by the monks of Westminster and discussed by Miss Harvey (*Westminster Abbey and its Estates*, pp. 246–54). Contractual tenancies, she says, made the greatest headway on those manors where villein dues still included labour rent on a significant scale down to 1348: it was on labour services that the leaseholders were turning their backs. This must surely have been the case at Glynde, too.

[125] The speedy return to normal was a theme of Miss Levett's work on the effects of the Black Death on the estates of St Albans Abbey (A. E. Levett, 'The Black Death on the St Albans Manors', in *Studies in Manorial History*, ed. H. M. Cam, M. Coate, L. S. Sutherland (reprinted London, 1962), pp. 253–5).

turned for relief to sheep grazing. In Sussex no such shift from one type of economy to another is observable. The county remained a predominantly arable one throughout the fourteenth century. Nevertheless, the pastoral element was of significance on those manors, like ours, which had access to the South Downs. In 1341 there were at least 110,000 sheep in Sussex as a whole, and the fact that Shoreham and Seaford were the main ports through which the local clip was exported suggests that the bulk of them—or certainly those bearing the finest and most highly-prized wool—were to be found in the eastern half of the county.[126] Some manors fielded very large flocks indeed. According to the Nonarum Inquisitions, which provided us with our figures for the total number of sheep, there were flocks of just over 3,000 in 1341 at Alciston, a Battle Abbey manor, and of 2,200 at Piddinghoe and Laughton.[127] It is unlikely that numbers were any fewer at the end of the century. As we shall see, flock sizes could fluctuate greatly from one year to the next, according to the incidence of disease.[128] But it would be difficult to establish any secular trend, either upwards or downwards, before about the mid-fifteenth century, when there may have been a shift in ownership from lords to peasants.[129]

Such speculations, however, take us beyond our present purpose, which is to measure the contribution made by sheep-rearing to the fortunes of our knightly proprietors. The stock inventories on the dorse of the accounts make the obvious starting-point. At Beddingham the number of head of sheep rose year by year in the brief period covered by our first four surviving accounts from 260 in 1307/8 to over 800 in 1310/11.[130] It is doubtful if the flock was being built up from scratch at a date as late as this; more likely it was being replenished after a visitation of the murrain. In the second half of the century it fluctuated between 700 and 1,000—that is, a couple of hundred lambs, 100–200 one-year-holds and then roughly equal numbers of ewes and wethers. At Glynde on the eve of the Black Death there were some 750 in all—probably the highest total ever

[126] R. A. Pelham, 'The Exportation of Wool from Sussex in the late-Thirteenth Century', *SAC* 74 (1933), 131–9, and idem, 'The Distribution of Sheep in Sussex in the Early-Fourteenth Century', *SAC* 75 (1934), 130–5.

[127] Pelham, 'The Distribution of Sheep', 1310–1; J. A. Brent, 'Alciston Manor in the Later Middle Ages', *SAC* 106 (1968), 90.

[128] See below, pp. 135–6.

[129] T. H. Lloyd, *The Movement of Wool Prices in Medieval England* (Economic History Review Supplements, 6, 1973), p. 28. This probably explains the apparent fall in numbers observed by Brent at Alciston ('Alciston Manor in the Later Middle Ages', Bristol MA thesis, p. 59).

[130] GLY 996.

reached there, and certainly never matched in the years in the 1360s and 1370s for which we have accounts.[131] At the Sackville manor of Claverham, where stock rearing was a speciality, we find quite large flocks again—nearly 1,000 in 1349/50, and large, though widely fluctuating, numbers in later years.[132] And at West Dean, another manor embracing large expanses of downland, we find nearly 890 head of sheep in 1387/8.[133]

Sheep farming was a large-scale business, therefore, and here as elsewhere it was organized on an inter-manorial basis. Beddingham, for example, seems to have been coupled with Peakdean, another Etchingham manor, a few miles to the south-east on the other side of the Downs. There the young lambs were driven from Beddingham in their first summer, and thence they were brought back twelve months later.[134] The clips of both manors were usually brought together at Beddingham for disposal to a buyer. There is a hint of specialization of function here: but it may be only a hint—for there was a great deal of movement between manors out of which no clear pattern seems to emerge. Claverham, for example, saw over 800 sheep either coming or going in the course of the twelve months 1365/6—293 lambs being received from Newland and 150 ewes from Chiddingly (a couple of miles to the north), and another 100 lambs and 100 wethers being driven on to Chiddingly. In addition, 63 lambs were bought and 119 others sold.[135] Whether it is possible to discern a policy of transhumance behind these and like movements and exchanges of animals is hard to say. In general, the hardier animals—the wethers, that is—were maintained at Chiddingly; but beyond that it would be unwise to go in the absence of a consecutive series of accounts.

However, exchanges and movements could sometimes take place between as well as within estates. In 1420 we find Sir John Pelham's stock-keeper at Laughton responsible for a vast flock of 4,000 sheep, some of them Sir John's, and others new arrivals from Jevington and

[131] In 1369/70 there were 19 wethers, 101 ewes, 37 two-year-olds and 119 lambs, giving a total of 276 (GLY 1074). In 1372/3 there were 20 wethers, 51 ewes, 7 two-year-olds, and 48 lambs, giving a total of only 126 (GLY 1076).

[132] In 1365/6 there were 119 wethers, 423 ewes, 300 two-year-olds and 945 lambs giving a total of 1,787 (SAS/CH 255), in 1379/80 115 wethers, 212 ewes, 231 two-year-olds, and 132 lambs giving a total of 680. But in 1380/1 there were only 91 two-year-olds and 9 lambs (SAS/CH 259, 260).

[133] SAS/M/673.

[134] This is the impression conveyed by the long series of consecutive accounts from 1361 to 1390 (GLY 998–1002).

[135] SAS/CH 255–61.

Birling, manors a few miles away held by other lords.[136] Why had they been entrusted to his custody? Had they been leased? Or were they there for shearing? These are questions we cannot answer. But Sir John's own sheep must surely have been heavily outnumbered by the others because back in the 1370s the demesne flock had never numbered more than 300.[137] In the light of this knowledge the figure of 2,200 sheep at Laughton given in the Nonarum Inquisitions takes on a new significance. Is it not possible that it may have included not just the demesne flock but those from a few neighbouring manors? And if so, how many other seigneurial stock-keepers looked after other flocks in addition to their own?

The very flexibility with which flocks were managed makes it all the more difficult to estimate how much they were worth to their owners.[138] The problem is this. In some years the clip was sold locally by the reeve, and in others centrally by the steward. If it was sold locally, as it was at Beddingham in 1310 to Roger de Ely and at Glynde in 1348 to John de Winchester, then the price was entered in the reeve's account.[139] If it was sold centrally, it was entered in some central account, now invariably lost. Our figures, culled only from reeves' accounts, form therefore a very irregular series.[140] They cover less than one year in two; but they do give a rough idea of how much the sale of wool might add to the profits of a manor. Beddingham

[136] BL Add. Roll 32152. Birling was owned by the Bardolf family and Jevington by the St Clairs.

[137] Clough, 'Estates of the Pelham Family', p. 74.

[138] They were kept, of course, primarily for their wool and only to a lesser extent for their meat.

[139] These two entries provide us with our only information about the purchasers of the wool. Roger de Ely was a resident of Seaford (GLY 1638). He represented his town in parliament in 1322 (T. W. Horsfield, *The History, Antiquities and Topography of the County of Sussex*, ii (London, 1835), Appendix p. 71). Winchester is harder to pin down. In fact, the statement that he was the purchaser is simply my reading of the entry in the account (GLY 1072) to the effect that it was he and Henry de Bainden, the reeve, who took the clip to Canterbury. Bainden, we know, was a local man. Winchester, on the other hand, was not. But he was someone who passed through from time to time. On 19 July 1347 he witnessed the grant of a piece of land from Sir John Waleys to one of his tenants, and his was one of the two seals used to authenticate the transaction (GLY 1165). Did he make a visit each year to pick up the clip? There was a John de Winchester involved in the wool trade in 1341 (*CCR 1341–3*, p. 169). Whether this was the same man as the John de Winchester who was a keeper of the peace in Hampshire and probably a resident of Southampton is, unfortunately, not clear (*CPR 1348–50*, pp. 382–3, 588).

[140] The view of account of the Sackville estates compiled after Sir Andrew's death in 1369 mentions sales of grain from a number of manors but, sadly, not sales of wool (SAS/CH 250).

usually accounted for a couple of sacks each year which, depending on the quality, and on the relative proportions of sheeps' and lambs' wool, would be sold for between £6 and £10; if the Peakdean clip was included, as it was in 1369, the profit was effectively doubled (see Table IV).[141] In other words, wool made a difference, perhaps even an appreciable difference, but did not transform the balance of profit and loss on a manor where the demesne arable was still the main source of income. The kind of property on which wool production could be practised most profitably was one on which arable cultivation was reduced to a minimum, allowing the economies of sheep-rearing to be exploited to the full. Such was the case at Claverham. The sums realized by the sale of wool from this manor, when we have them, fluctuated wildly—from £5. 17*s*. 0*d*. in 1354/5 to no less than £41. 8*s*. 9*d*. in 1366/7, when the wool and wool-fells of that and the previous year were, unusually, sold together. However, the net profits here would have been higher than elsewhere because fewer staff had to be employed. A couple of shepherds and a cowherd, that was all. Throw in £1 for the maintenance of the sheepfold, and the running costs still came to less than £5 a year.[142]

Are we entitled to conclude, then, that the sheep were indeed the panacea they are made out to have been? That if the trappings of an obsolescent demesne husbandry had been swept away, allowing sheep grazing to be practised with maximum economy, then landowners could have realized profits they would never have dreamed of? Not quite. Medieval sheep were notoriously prone to disease—to the host of ailments collectively described in the accounts as 'murrain'—and the cost of replacing a flock could wipe out all the previous years' profits. The lowland sheep seem to have suffered far worse than those pastured on the Downs. The mortality rates at Beddingham, Heighton and West Dean, for example, were not high.[143] But at Claverham they were sometimes calamitous. In 1366/7, when there was a nation-wide outbreak of murrain, 110 of the 367 ewes, 126 of the

[141] Although it is clear from the accounts that the Peakdean clip was often sold with that of Beddingham, only in 1368/9 was it sold locally and the price entered in the Beddingham account.

[142] Expenses at Claverham came to £4. 8*s*. 2*d*. in 1354/5, £15. 19*s*. 0*d*. in 1365–6 (when cows and calfs were purchased), £5. 12*s*. 1*d*. in 1366/7, £4. 7*s*. 2*d*. in 1379/80 and £4. 19*s*. 2*d*. in 1380/1 (SAS/CH 252–61). In each case I have included 'vadia' payments, which were always charged to the Chalvington account, as explained above, p. 00.

[143] At Beddingham the only years that stand out are 1375/6, when 6 wethers, 11 ewes, 20 2-year-olds, and 106 lambs died, and 1390/1 when 86 lambs died.

368 two-year-olds and no fewer than 401 of the 931 lambs died.[144] Still, the flock was built up again, though how quickly we do not know, and there were some 600 sheep there by the end of the next decade. But then the murrain struck again. In 1379 all but 6 of the wethers died and, worse still, all but 2 of the ewes, and that before lambing.[145] The flock was decimated. And it could not be built up again through breeding. Two years later there were 91 2-year-olds and 9 lambs, and still only 2 ewes.[146] Restocking was going to take time—and money.

The old adage about the danger of putting all one's eggs in a single basket had some relevance to sheep farming at Claverham. It had its attractions, of course. Running costs could be reduced to a minimum. But it was still a risky business. In some years the receipts might be pure profit; in others they might be swallowed up by the cost of re-stocking. Wool production was not the godsend it is sometimes made out to have been. Far more dependable, in fact, as a source of income at Claverham was the dairy which, whether farmed or managed directly, was never worth less than £8. 16*s*. 0*d*. or more than £13. 13*s*. 10*d*. in the years covered by the surviving accounts.[147] For a herd of some 29 cows at the least and 45 at the most that was not a bad level of income.

On other manors, by contrast, the demesne livestock were out-numbered by those of the peasants. At Udimore, for example, Sir William de Etchingham V rarely had more than the eight oxen needed to pull the ploughs, a few horses and a couple of dozen pigs. Rather than graze his own animals he chose to profit by charging his tenants for the right to graze theirs on his land. The amount of land for sale varied from year to year, but it must always have been a fair acreage, and in every year for which we have accounts it realized more money than did sales of demesne produce.[148] Receipts rose to a peak of £19. 16*s*. 0*d*. in 1366/7, but then fell back to £10 at the beginning of the next decade. It was apparent that Sir William would not always be able

[144] SAS/CH 256. For murrain outbreaks in the middle ages see Lloyd, *Movement of Wool Prices*, pp. 14–20.

[145] SAS/CH 259.

[146] SAS/CH 261. For a more optimistic view of the possibilities of sheep farming see C. Dyer, *Lords and Peasants in a Changing Society: The Estates of the Bishopric of Worcester, 680–1450* (Cambridge, 1980), pp. 139–40.

[147] SAS/CH 252–61.

[148] The only exception is 1370/1, when exceptionally higher receipts of £14. 4*s*. 2*d*. took sales of grain to £18. 18*s*. 9*d*. That same year income from pasture had fallen to £10. 11*s*. 2*d*. (Hastings Museum, JER/Box 3).

Table IX Udimore: Sales of Pasture Rights (some small fields are omitted)

	1361/2	1362/3	1363/4	1364/5	1365/6	1366/7	1368/9	1369/70	1370/1	1380/1
Billingham winter	17*s.* 6*d.*	£1 2*s.* 0*d.*	15*s.* 0*d.*	15*s.* 0*d.*	15*s.* 0*d.*	15*s.* 0*d.*	£1 05*s.* 9*d.*	14*s.* 2*d.*	16*s.* 0*d.*	let at
Billingham summer	£4 0*s.* 0*d.*	£4 0*s.* 0*d.*	£4 0*s.* 0*d.*	£4 13*s.* 4*d.*	£4 3*s.* 4*d.*	£4 2*s.* 0*d.*	£5 11*s.* 1*d.*	£4 3*s.* 6*d.*	£4 4*s.* 0*d.*	farm
Creggesmarsh	19*s.* 0*d.*	£1 0*s.* 0*d.*	£1 0*s.* 0*d.*	£1 0*s.* 0*d.*	8*s.* 0*d.* (partly animal fodder)	18*s.* 0*d.*				
Conyer and Pipenesell	£1 0*s.* 0*d.*	£1 0*s.* 0*d.*	at rent	at rent	12*s.* 0*d.*	15*s.* 0*d.*	11*s.* 6*d.*	let at farm	let at farm	let at farm
La Pyke and La Broke*	10*s.* 8*d.*	13*s.* 4*d.*	16*s.* 0*d.*	£1 14*s.* 0*d.*	£1 12*s.* 0*d.*	£1 12*s.* 0*d.*		£1 2*s.* 1*d.*	1*s.* 10*d.* (partly seeded)	
Birchesgate and La Marlyng	12*s.* 0*d.*	£1 5*s.* 0*d.*	£1 6*s.* 0*d.*	8*s.* 0*d.* (partly seeded)	8*s.* 0*d.*	8*s.* 0*d.*	7*s.* 0*d.*			
Longmarlyng	5*s.* 0*d.*	6*s.* 0*d.*	7*s.* 0*d.*	9*s.* 0*d.*	under seed		1*s.* 0*d.*		5*s.* 6*d.*	let at farm
La Damme	under seed	under seed	under seed	under seed	meadow	meadow			5*s.* 0*d.*	let at farm
Wymondsland and Smythsland					9*s.* 6*d.*	9*s.* 6*d.*	10*s.* 0*d.*	let at farm	let at farm	let at farm
Cow pasture	£4 0*s.* 0*d.*	£3 7*s.* 0*d.*	£4 16*s.* 8*d.*	£5 13*s.* 4*d.*	£5 0*s.* 0*d.*	£4 0*s.* 0*d.*	£4 0*s.* 0*d.*	£4 0*s.* 0*d.*	£4 0*s.* 0*d.*	
Agistment in park						£5 0*s.* 0*d.*				
Regges						14*s.* 0*d.*	8*s.* 0*d.*			

* entered under Billingham

to ask as much as he had in the mid-1360s. For winter pasturage at Billingham, for example, he charged 17*s*. 6*d*. in 1361/2, £1. 2*s*. 0*d*. in 1362/3 and only 15*s*. thereafter. He had to lower his sights. Again, he charged 12*s*. for Birchesgate and La Marlyng in 1361/2, £1. 5*s*. 0*d*. and £1. 6*s*. 0*d*. for the next two years and thereafter 8*s*. for the part that was still pasture (the remainder now being seeded). It is possible that these fluctuations in income reflect not market pressures but changes in the relative proportions of grass and corn. This is possible, certainly. But if there was a change in land use, that was usually noted in the account. It seems on balance more likely that Sir William aimed to charge as much as he could, and if it was too much one year he lowered his sights the next.

It may also be objected that these figures can no more be trusted than those for wages and stipends which we have already considered: that, like those, they have become fossilized, and therefore mislead us more than they inform. But scepticism can be taken too far: the small but significant variations in the figures suggest that they do actually represent the sums that were charged. But whether they do or do not, the fact that it is so difficult to tell imposes serious limitations on the use that can be made of the accounts as a whole. They cannot be used to measure the movement of wages. Nor can they be trusted to show how much was collected in rents. What they tell us is not how much the lord received from his lands but how much he received from his reeve. For the historian wishing to trace the changing fortunes of the gentry, however, that information is just as important. The lord's spending power, after all, was determined not by the source of his income but by its level. And there the accounts leave us in no doubt. In the late 1370s it was dropping.

On the manors we have studied that drop—that quite sharp drop—does not seem to have occasioned any correspondingly sharp adjustments in the way the manor was managed: for none was necessary. The relationship between the lord and his local manager, the reeve, had already begun to change under the impact of the rise in wages after the Black Death. The reluctance of proprietors to sanction wage increases on the scale demanded by rural labourers left the reeve with no alternative but to pay them himself and to recoup his losses elsewhere—partly no doubt at the expense of the tenants and of others with whom he had dealings, but partly too at the expense of the lord. Proprietorial acquiescence in this arrangement was marked by a relaxation in the rigour of accounting.

If we can see how the knights adjusted their management techniques to movements of prices and wages, it is more difficult to see how they adapted their style of life. Did they make economies? Did they go on spending as freely as before, hoping that costs would soon fall again? Or did they seek ways of increasing their income? On the evidence we have before us we cannot say. We can only speculate, and the speculations cannot take us very far. We do not know, for example, whether any of our knights took leases of other lords' demesnes.[149] Nor do we know whether, or how far, they encouraged changes in land use on their own estates. On the evidence of the accounts the answer is probably very little. But it would be unwise entirely to rule out the possibility. There are signs that lords of this rank (or their officials) knew that advantages could be gained from specialization. Chalvington and Claverham, as we have seen, were run as two complementary economies, the former specializing in corn-growing, the latter in sheep-grazing. These two manors lent themselves to such management, because they lay adjacent to each other, and after the 1330s were in the ownership of the same family. Specialization was harder to practise on an estate the constituent units of which lay miles apart. But it could be done. The taking of demesne leases offered one obvious way of rearranging the estate geographically. When Sir Roger Ashburnham leased from William Batsford 117 acres of arable land, 51 acres of thicket and marsh and 35 acres of wood in Ewhurst immediately to the east of his own lands, he was taking the first of a series of steps which were eventually to make him and his descendants the dominant lords in that village.[150] The leasing of sheep flocks offered similar advantages in respect of stock-grazing—advantages of which, as we have seen, Sir John Pelham for one availed himself when he took in his neighbours' flocks. If these examples are but two of many such arrangements that proprietors made with each other, then a combination of rearrangement of holdings and flexibility in land use could have opened up opportunities scarcely hinted at in the accounts we have discussed.

[149] One instance from another family is noted below.
[150] *CCR 1389–92*, p. 76.

V

ETCHINGHAM CHURCH

ETCHINGHAM CHURCH amply justifies the praise that has been lavished upon it. To find a fourteenth-century church so little altered, so little touched by the hand of the restorer, is rare enough, to find one of this distinction even rarer. 'This grand church', wrote the author of the *Little Guide* in 1949, 'is easily the finest of its kind in Sussex.'[1] 'Grand' is an apt description. Sir Nikolaus Pevsner preferred 'large and proud'.[2] But neither writer felt tempted to say 'beautiful', and with good reason. Curiously perhaps for a church of the 1360s, Etchingham no more anticipates the elegance which was to characterize Perpendicular than it retains any of the delicacy that had marked the Decorated style. It is a sturdy building, solidly masculine in quality, and reliant for its aesthetic effect not on linear richness or surface texture but on a bold grouping of masses. The design is ambitious, and the proportions cathedralesque. The tower is placed not at the west end, but centrally, linking a short but high nave of two bays with a longer chancel of three. A lofty clerestory rises above the nave aisles which are themselves continued eastwards on each side of the tower to form chantry chapels. For a church of the second half of the fourteenth century the positioning of the tower is a little anachronistic. It may perhaps be that the plan followed that of an earlier church on the site. But the impression that for its date this is an old-fashioned church is strengthened by a glance at the window tracery. It has all the intricacy, all the richness of variety, of curvilinear tracery of the 1330s; yet it could not have been installed any earlier than the 1360s.

In the absence of any fabric rolls we cannot say for certain when Etchingham Church was begun or completed; but we can say roughly when work was in progress and who was paying for it. In 1368 Sir William de Etchingham claimed in Common Pleas that Thomas Esshyng, a mason of Betchworth (Surrey), had failed to honour a contract, made in 1363, to instal five windows each of three lights in

[1] R. F. Jessup, *Sussex* (The Little Guides: London, 1949), p. 139.
[2] I. Nairn and N. Pevsner, *Sussex* (Harmondsworth, 1965), p. 497.

Etchingham Church.[3] After several adjournments the case disappeared from the rolls in Michaelmas the following year, doubtless settled (like so many others) out of court. Whether Sir William retained Esshyng's services we do not know, but it seems unlikely. The windows eventually constructed have two lights, not three, and the stone employed, as in the rest of the building, is Hastings sandstone and not the local Reigate stone in which a Surrey mason might be expected to have worked. The chances are that Sir William turned to someone else. For our purposes, what the case places beyond doubt is that this church was the product of seigneurial, not communal, patronage. The lord hired the masons and scrutinized the plans. And the pride that he took in their execution is made manifest in the inscription on his brass (Plate 1), which removes any lingering doubt about where the responsibility for rebuilding lay:

> Iste Willelmus fecit istam ecclesiam de novo
> reedificari in honorem Dei et Assumpcionis Beate
> Marie et Sancti Nicholai . . .

Sir William died on 18 January 1389—'entour mynoet', about midnight, as the foot inscription delightfully tells us—and was laid to rest in the new church which, we can take it, would by then have been completed. He was probably the first of his line to be buried there. The point is difficult to substantiate, because our knowledge of the burial places of his ancestors is far from complete. But what evidence there is leaves little room for doubt. No family tombs were transferred from the old church at Etchingham to the new, while Sir William was heir to a strong family link with the two local monasteries of Battle and Robertsbridge; and those of his ancestors whose burial places are known had opted to be interred in the one or the other.

The Etchinghams and the monks of Battle both traced their origins to the earliest days of the Norman settlement, the Etchinghams to their ancestor Reinbert, steward of the count of Eu and first sheriff of the rape of Hastings, and the monks of course to William the Conqueror's vow to found a monastery on the site of his great victory.[4] As (so they claimed[5]) hereditary sheriffs of the rape, successive

[3] PRO, CP40/430 m. 174d, noted by J. Harvey, *English Medieval Architects: a Biographical Dictionary down to 1540* (London, 1954), p. 101.

[4] For the role of the early Etchinghams in the rape of Hastings see Searle, *Lordship and Community*, pp. 51–3, 210–11.

[5] It was contested in 1207 (S. Hall, *Echyngham of Echyngham* (1850), p. 2).

generations of the Etchingham family doubtless saw much of the monks, and from time to time made grants to them. Simon de Etchingham (died *c*.1244) granted them an annual rent of 3½ marks in Whatlington, reserving to himself the presentation to the chapel there.[6] His son William enlarged the grant, and chose to be buried in the abbey; and *his* son in turn, another William, finally parted with the fief altogether.[7]

How much we are entitled to read into these gifts is hard to say. The transfer of an advowson was after all a cheap way of securing salvation by good works at the same time as bowing to reformist disapproval of secular ownership of benefices. The evidence suggests that by the late thirteenth century family favour was being channelled more in the direction of Robertsbridge. This abbey had been founded in 1176 by one Alured of St Martin on a site close to the Etchingham's manor of Salehurst. In the early thirteenth century its somewhat meagre endowment was augmented by grants from the family of the counts of Eu, to whom the right of patronage had passed; but 100 years later, when financial disaster threatened, it was the Etchinghams and their circle who came to the rescue. Since the 1280s, if not before, the abbey's estates had suffered badly from inundations by the sea, and in 1309 Sir William de Etchingham IV proposed to make good the consequent loss of revenue by alienating the three churches of Salehurst, Udimore and Mountfield, reckoned together to be worth 50 marks per annum.[8] Five years later, in 1314, he granted the community lands in Salehurst for the purpose of supporting two chaplains to perform daily services for the safe repose of the souls of himself, his wife, and his heirs. It was almost certainly, too, at his expense that the abbey church itself was enlarged in the early fourteenth century. His wife and daughter were both laid to rest there;[9] and the provision of a grander eastern arm for the monks could have been his way of honouring their memory. Sir William IV, then, was by far the most important benefactor of the abbey in the early 1300s; but by his example he inspired others, notably his friend Sir Alan Buxhill, to make their own, albeit smaller, grants to the community.[10]

[6] *VCH Sussex*, ix. 112–13.

[7] ibid. By the later fourteenth century relations with Battle had cooled. In the 1360s the abbot was suing Sir William de Etchingham V for the performance of services exacted by the lord of the rape, John of Gaunt (PRO, CP40/430 m. 294d).

[8] *Cal. Robertsbridge Charters*, no. 290; *CPR 1307–13*, p. 159.

[9] *Cal. Robertsbridge Charters*, nos. 300, 321.

[10] Ibid., nos. 288, 298.

By the third and fourth decades of the fourteenth century donations to the monks were beginning to fall off, at Robertsbridge no less than elsewhere. No later members of the Etchingham family were to show the generosity that Sir William IV had, and such gifts as the community received after 1330 from other patrons were both fewer and individually less valuable than those they had received before. Piety was evidently finding an outlet in other directions, in the case of Sir William V in the next generation in the rebuilding of the parish church. This redirection of patronage was not, of course, unique to the Etchingham family. It may be going too far to say that it was a general phenomenon among the upper classes of late-medieval England, but it has been observed often enough elsewhere to call for a word of explanation. If we look at the Berkeleys of Coberley (Gloucs.), for example, we find that Sir Giles who died in 1294 wished to be buried at Little Malvern Priory (Worcs.) and that his son and grandson both opted for the village church at Coberley. Over roughly the same number of generations their neighbours and noble kinsmen, the Berkeleys of Berkeley Castle, who had usually been buried in St Augustine's Abbey, Bristol, of which they were patrons, began to show a preference for burial instead in the parish churches of their manors of Berkeley and Wotton-under-Edge.[11] These were two families which had long enjoyed rights of patronage over a monastery, and might therefore be expected to have sought burial within its walls. But those who had no such a claim on a house, it might be argued, would very likely have been buried in their village churches all along. Perhaps they were; but the point is difficult to prove because all too often the sepulchral evidence we need was swept away in rebuildings of the fourteenth and fifteenth centuries—rebuildings which themselves attest the increasing level of interest shown in parochial fabrics in those two centuries.

In east Sussex no fewer than three churches were completely rebuilt within a few years of each other in the later fourteenth century—Alfriston, in the Cuckmere Valley, Poynings, half-way along the Downs, and of course Etchingham. In each case work had probably been begun before the decline in incomes in the late 1370s put paid, at least for the time being, to extravagant building programmes. Dating

[11] N. E. Saul, 'The Religious Sympathies of the Gentry in Gloucestershire, 1200–1500', *Trans. Bristol & Gloucs. Archaeological Soc.* xcviii (1981), 103–4. See also M. G. A. Vale, *Piety, Charity and Literacy among the Yorkshire Gentry, 1370–1480* (Borthwick Papers, 50, 1976).

is, of course, notoriously difficult. Yet in this case it is also important, because the similarity in conception of these three churches suggests that one might have provided the model for the other two. Each is more or less cruciform, and each, unusually for the period, has a central tower. For Alfriston there is only the evidence of the building itself to guide us, and that looks more Perpendicular than Decorated.[12] It may therefore have been begun a year or two later than Etchingham. For Poynings, however, there is the evidence of Sir Michael de Poynings's will, made in September 1368, which provided for 200 marks to be spent on rebuilding the church there. Sir Michael died in March 1369, and work may be presumed to have begun in about 1370—or even a year or two later, because this is a more purely Perpendicular church than either of the other two.[13]

The implication, therefore, is that of the three churches Etchingham was the earliest. Whatever the reasons that may have led Sir William to rebuild the place, it was not a desire to keep up with the fourteenth-century Joneses. He was playing the part of Mr Jones himself. He was setting the fashion, not following it.[14] In that case, what did lead him to think of rebuilding, and to do so at a time when the wages of building labourers were rising a good deal faster than those of estate labourers?[15] It is not as if his family were lacking a mausoleum or burial place. As we have seen, they had one at Robertsbridge Abbey. Sir William, however, was not satisfied with following the habits of his ancestors. Not for him the notion of 'mos maiorum'. But why? What, if anything, had gone wrong?

Relations between the family and the monks of Robertsbridge, as we have seen, reached their lowest ebb in the early 1330s when Simon, the rector of Herstmonceux, challenged the community's appropria-

[12] Pevsner says that is is 'on the boundary between Decorated and Perpendicular . . . Some windows still decorated (two lights) others purely Perpendicular (with panel tracery)' (*Sussex*, p. 396). Alfriston is undoubtedly an impressive church. It is all the more frustrating, then, that there should be no evidence to suggest who built it.

[13] *Testamenta Vetusta*, ed. N. H. Nicolas (London, 2 vols. 1826), i. 73; Pevsner, *Sussex*, p. 586.

[14] It may appear inconsistent to portray Sir William de Etchingham on the one hand as a pace-setter in the rebuilding of churches and on the other as a conservative in matters of style. But Sir William did not design the church himself. That would have been the responsibility of his master mason.

[15] J. L. Bolton, *The Medieval English Economy, 1150–1500* (London, 1980), p. 274; M. M. Postan, *Essays on Medieval Agriculture and General Problems of the Medieval Economy* (Cambridge, 1973), pp. 199–200. The building of three substantial churches at roughly the same time would, moreover, have had the effect of bidding up building wages in this area—not to mention the work on fortifications to fend off the French.

tion of the rectory of Salehurst. His action, brought by writ of *quare impedit*, was unsuccessful; but, as the surviving narrative shows, it provided the opportunity for other, more powerful foes to join in the attack.[16] The first to do so was Robert de Tawton, the keeper of the privy seal and an erstwhile ally of Simon's, who brought his own writs of *quare impedit*, and challenged the validity of the patents of appropriation. He was soon followed by the chief justice, Sir Geoffrey le Scrope, who acted on the somewhat belated discovery that the disputed church formed a prebend in the king's free chapel in Hastings Castle. The abbot, summoned to defend the action, entered a demurrer, and used the breathing space that he had won to pick off each of his opponents in turn. To Simon he gave a present of 20 marks, later increased to 30. To Tawton he granted an annuity of 80 marks. And to the chief justice he offered an amended version of the composition agreed between himself and the dean of the free chapel. By February 1334, then, the way had finally been cleared for the abbot to be installed in his prebendal stall in the chapel, in the presence of Simon de Etchingham and many other local notables.

As far as we know, that ceremony brought to an end the rift in relations between the Etchinghams and the monks of Robertsbridge. In the next generation Sir William V resumed his family's tradition of support for the community. He responded to the difficulty they experienced in maintaining observances at the pre-Black Death level by agreeing to release them from the obligation imposed by William IV of finding two chaplains to perform daily service in the chapel built on the spot where the abbey had first been founded.[17] But while willing to relieve them of burdens imposed by his ancestors, he showed no inclination to bestow on them any new favours of his own. It was not that he bore them any ill will: at a social level his relations with them were probably good.[18] It was simply that, after the fashion of the time, he chose to make the parish church rather than the monastery the focus of the religious, and to some extent the social, aspirations of his family.[19]

[16] BL Add. MS 28550. See above, pp. 94–7.

[17] *Cal. Robertsbridge Charters*, no. 362.

[18] A point stressed by Saul, 'The Religious Sympathies of the Gentry in Gloucestershire', 109. The strength of the social and political ties that existed in the late middle ages between the monks and their gentry neighbours is sometimes evidenced by displays of heraldry on the outside of abbey gatehouses. The famous one is Butley Priory, Suffolk, illustrated in J. Evans, *English Art, 1307–1461* (Oxford, 1949), plate 35b.

[19] I have to choose my words carefully. Religious and social aspirations overlapped:

If the idea of enhancing the dignity of the parish church was therefore Sir William's,[20] the immediate justification for so doing was provided by the needs of his tenants. The earliest church at Etchingham was probably a chapelry of Salehurst. It lacked burial rights of its own (a sure sign of dependent status), and the villagers were obliged to carry their dead to the mother church, three miles to the south-east.[21] This, of course, they found irksome, and in the wake of the mortality caused by the Black Death they petitioned for a cemetary to be set aside for their use at Etchingham. In February 1357 Pope Innocent VI wrote to the bishop of Chichester asking for this to be arranged, and five years later, in 1362, Sir William de Etchingham obtained a licence for the alienation to the parson of the single acre of land that was required.[22] Now this, of course, brings us to the very time that the rebuilding got under way, and it may be possible to connect Sir William's decision to embark on this project with the recent recognition of the church's independence. If so, he may have been responding to a growing awareness of the importance of the parochial unit in the life of the Church and to an acceptance by the laity of some share of responsibility for the upkeep of the fabric.

Such ideas were hardly new; but they had been given clearer articulation in the thirteenth century. At the Fourth Lateran Council summoned by Pope Innocent III in 1215 some 60 or 70 canons had been agreed with the general aim of improving the quality of cure of souls. As the opening words of c. 27 put it, the government of souls is

they certainly cannot be separated in analysing the reasons that led a gentry family to rebuild their local parish church.

[20] Or that of his circle.

[21] I am grateful to Dr John Blair for advice on the subject of early minsters. The first reference to the church comes in the *Taxation Ecclesiastica Anglie et Wallie Auctoritate Papae Nicholai IV circa 1291* (Record Commission, 1802), p. 137, where it is valued at £8. Salehurst with Udimore, by contrast, was valued at £24.

[22] *Cal. Papal Registers, iii, 1342–62*, p. 583; *CPR 1361–4*, p. 224. The implication of the licence is that it was demesne land that was being alienated. The manor-house is indeed believed to have stood a little to the east of the church, approximately on the site of the present-day railway station and coal-yard. According to *The Manor of Etchingham cum Salehurst*, ed. Sir S. P. Vivian (SRS liii. 1953), xxvii, the moat surrounding the manor was extended westwards, probably at the time of the rebuilding, so as to include the church; but no evidence is cited in support of this assertion. To imagine the visual impression that the new church would have made, it has to be remembered that it and the manor-house stood virtually alone. There was no nucleated village. Etchingham in the middle ages was a parish of scattered settlements. The modern village, I suspect, is a product of the railway age. It has few if any vernacular dwellings of pre-nineteenth-century date, in marked contrast to Burwash, a couple of miles higher up in the Weald.

the art of arts. The intention, to quote Powicke, had been 'to provide a body of disciplined, educated clergy, armed with an orthodox creed against the heretical tendencies of the time, and qualified, by character and training, to instruct their parishioners and to hear their confessions at least once a year.'[23] Innocent expected metropolitans to continue the work begun by the Council by holding provincial assemblies of their own. In England only two such were ever held in the thirteenth century, in 1222 and 1279–81, but most bishops did summon diocesan synods at which statutes modelled on those of the Councils were published, with the purpose of correcting widespread abuses and laying down future guidelines for the direction of the clergy. Standards of pastoral care may not always have been as bad as the reformers sought to portray, but all the same much remained to be done. The problem was that the battles fought back in the twelfth century to emancipate the church from what had been seen as the corrupting influence of secular lords and princes had actually done little to advance the cause of saving souls. Ironically, one reason for this lay in the very success of the monastic wing of the reform movement. The monks had been liberally endowed with churches, out of the revenues of which they would pay, often inadequately, a vicar. Whether the parish was served by an absentee pluralist rector or a poorly paid, ill-educated vicar, the end result was often much the same: cure of souls was neglected. The consequence was a growing concern in the thirteenth century with parochial reform. This found expression in a number of ways, not all of them directly concerned with the parochial organization. For example, the friars, who technically lay outside its framework, were brought in to preach and to minister to the faithful in areas ill-served by the parochial system. Clearly, there was a great appetite for preaching in the thirteenth century, as great perhaps as in the seventeenth, and parsons were frequently enjoined to mount their pulpits more often.[24] Other measures, however, focused more directly on the parish church. There were the growing demands made on the laity to provide fittings and furnishings for the church and to accept a measure of responsibility for its upkeep. It seems possible that at first the authorities, somewhat cheekily, tried to pass on to them the burden of maintaining the chancel as well as the nave, but by the late thirteenth century it was fairly

[23] F. M. Powicke, *The Thirteenth Century* (Oxford, 2nd edn., 1962), p. 449.

[24] Perhaps it is no coincidence, then, that the earliest surviving pulpits in England date from the fourteenth century.

generally accepted that the parson was responsible for the former and the laity only for the latter.[25] As the demands made grew in number, and doubtless in intensity too, so parishioners thought to form themselves into vestry associations and, if they held sufficient property, to appoint churchwardens to act as custodians. There is evidence as well that groups of parishioners, whether or not they formally constituted themselves as vestries, would contract to undertake repairs to the fabric.[26] Small expenditure on this scale would be within their means and competence. But if a total reconstruction were required, the patron and lord of the manor would be the person to whom they would look to foot the bill. It was this responsibility that Sir William was shouldering at Etchingham in the 1360s.

For Etchingham is above all Sir William's church. It is not, as so many English churches are, a building that has evolved slowly over the centuries in response to changing liturgical fashion or the needs of a growing population. It is the product of a single building programme carried out in the behest of this one man, and bearing every sign of his personal involvement. The most striking manifestation of this was once to be found in the scheme of heraldic decoration conceived to fill the stained glass windows. Sadly, here as elsewhere, time has taken a heavy toll of this most vulnerable part of the medieval heritage, and what little remains is now chaotically re-set in the windows of the nave.[27] Its original character can be savoured from a set of drawings made by Grimm in 1784 which showed that each of the two main lights contained in the centre a saint under a canopy, at the foot a square panel with a coat of arms and in the tracery one or two more coats of arms. Grimm's record of the heraldry is, however, sometimes at variance with that of William Hayley, who had taken copious notes when visiting the church six years earlier in 1778. In view of the latter's general reliability, it is his testimony that has been followed here.[28]

In the east window were the arms of Edward III, the Black Prince, John of Gaunt, duke of Lancaster, and John, duke of Brittany, the king's son-in-law and ally. Next on each side came the arms of the rebuilder himself and perhaps of his son, and in the remaining chancel

[25] E. Mason, 'The Role of the English Parishioner, 1100–1500', *Jnl. Eccles. Hist.* xxvii (1976), 24–5.

[26] Ibid. 24–6; C. Platt, *The Parish Churches of Medieval England* (London, 1981), pp. 88–97.

[27] For a description see F. Lambarde, 'Coats of Arms in Sussex Churches', *SAC* lxvii (1926), 151–2.

windows those of the English earls. The glass in the nave was already too mutilated by Hayley's time to permit a complete reconstruction, but it is reasonably clear that the coats belonged to gentry families of east Sussex.

What we need to establish are the considerations that influenced Sir William firstly in the selection of these coats of arms and secondly in their arrangement in the windows. How did he decide whose coats of arms to include and whose to omit? It is understandable that he should have wished to give pride of place to the king and the heir to the throne in the east window. But what about the earls? Was he to include them all or just a select few? The number of coats recorded by Hayley is simply too great to admit of the possibility that he honoured only those under whom he had fought or by whom he had been retained. Besides, there is little surviving evidence to show that he had seen much active service on the battlefields of France and none at all to suggest that he had ever been retained by indenture.[29] No, the intention, as Hayley discerned two centuries ago, must have been to include them all. In that case, a *terminus a quo* for the glazing of the chancel windows is provided by the inclusion of the arms of John of Brittany, earl of Richmond, to whom the rape of Hastings was granted in 1372, and a *terminus ad quem* by the omission of those of the earls of Buckingham, Northumberland, Nottingham, and Huntingdon, who were not to receive their titles until Richard II's coronation in 1377.[30] We are given a date of roughly the mid-1370s, which would be consistent with what we have seen of the pace of work on the erection of the church. As for the coats of arms in the nave, Hayley's notes give the impression that they belonged not only to relatives but to friends and neighbours of Sir William. Northwode, Dallingridge, St Clair, Shoyswell, and Waleys are the arms which he could distinguish and which are still in the church today.[31]

28 BL Add. MS 6344, fos. 185^r–185^v.

29 On the evidence of letters of attorney William seems to have fought in only two campaigns, in 1355 in the company of Sir Michael de Poynings (PRO, C76/33 m. 10) and in 1359–60 in that of Sir John de Crioll (PRO, C76/37 m. 1). I am grateful to Andrew Ayton for the latter reference.

30 For the circumstances surrounding the grant of the rape (which went with the earldom of Richmond) to John, duke of Brittany see J. J. N. Palmer, *England, France and Christendom, 1377–99* (London, 1972), pp. 18–19.

31 With the possible exception of Shoyswell, these seem to have been acquaintances rather than kinsmen. But, as we have stressed above (p. 7), before the end of the fourteenth century we can be none too sure who the Etchinghams' kinsmen were.

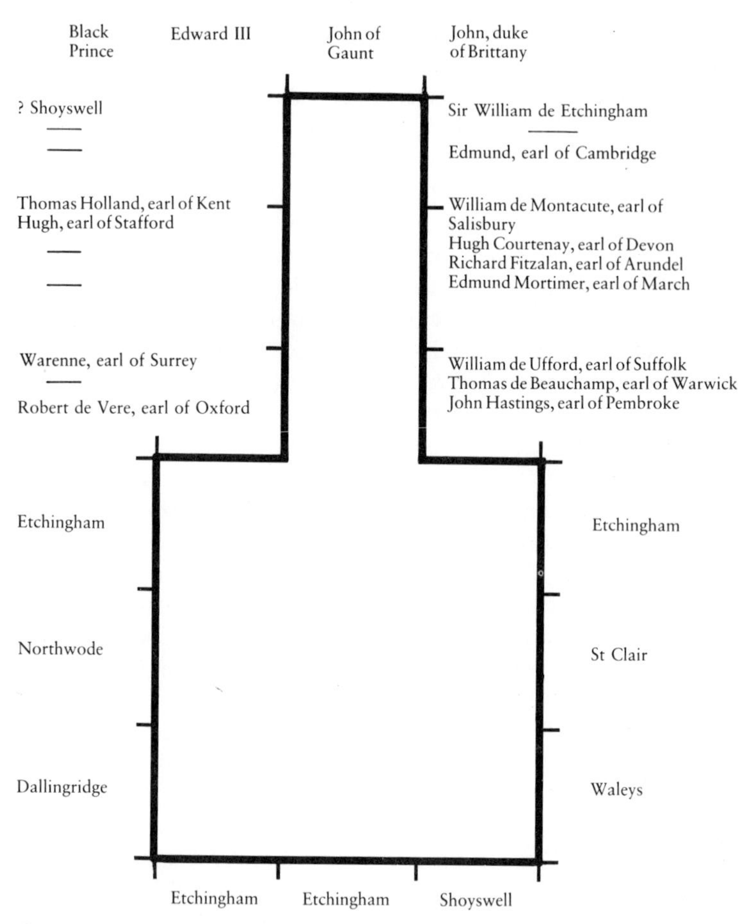

Etchingham Church—Plan of the Stained Glass Windows

Schemes of heraldic decoration were of course to become fashionable, not to say ubiquitous, in the late middle ages, in ecclesiastical settings as much as secular. They answered the need felt by a preliterate society for a form of identification that would both proclaim genealogical or political connections and commemorate the generosity of a benefactor.[32] The power of connections is evidenced strongly

[32] The fashion was probably set by the series of arms in the spandrels of the wall arcades of Henry III's Westminster Abbey, for which see P. Brieger, *English Art, 1216–1307* (Oxford, 1957), pp. 121–2. These are often said to be the arms of benefactors, but David Carpenter tells me that it is doubtful if they actually are: they represent most, if

in the heraldic scheme at Etchingham—but connections of no more than a very general kind. There is no evidence that any of the earls or knights whose arms were included had contributed to the cost of the church. Nor, as we have seen, is there any evidence that any of them had retained Sir William by indenture. The coats of arms cannot in that case be taken to be emblematic of a 'bastard feudal' relationship.[33] Rather they seem to convey a sense of delight in the use of heraldry for heraldry's sake. They proclaim in the language of visual symbolism the solidarity of all the noble- and gentle-born, the sense of pride, perhaps even the sense of separateness, felt by those of armigerous rank. But within the assembled company Sir William recognized a clear order of precedence. In the east window he placed the arms of the king and members of the royal family, and in the windows flanking them those of the earls. The arms of his knightly neighbours in Sussex he relegated to the nave. In other words, he conceived of the earls as comprising an estate of their own, recognizing perhaps the developments over the previous half century or more that had led them to draw apart from the knights.[34] To say that is not to deny that Sir William wanted to bask in the reflected spendour of his superiors. How else could one account for the inclusion of his own family's coat of arms in the easternmost windows of the chancel with those of the

not all, of the top nobility (including de Montfort), not just those connected with the rebuilding. Another fine series of arms, this one fourteenth century in date, is to be seen in York Minster running the whole length of the building in the spandrels of the main arcade; again, the range of families represented is wide, too wide, one feels, to admit in every case of a direct connection with the construction of the Minster. They are listed in A. Clutton-Brock, *York* (Bell's Cathedral Series: London, 1899), pp. 75, 118. To sound a note of scepticism about these very grandest of heraldic displays is not to deny, however, that in many other instances the presence of a coat of arms may well commemorate a patron's generosity to a church. It would be idle to deny, for example, a connection between the incorporation of the arms of Clare and Berkeley in the reredos of the lady chapel of St Augustine's Abbey, Bristol, and the rebuilding of that chapel in the early fourteenth century. The Berkeleys were patrons of the community, and the de Clare earls of Gloucester prominent landowners in the area.

[33] Unlike the badges of the late-medieval and early-Tudor period which David Starkey discusses so illuminatingly in 'Ightham Mote: Politics and Architecture in Early-Tudor England', *Archaeologia*, 107 (1982), 153–63, and idem, 'The Age of the Household', in *The Late Middle Ages*, ed. S. Medcalf (London, 1981), pp. 270–4.

[34] It was in the first half of the fourteenth century that notions of 'peerage' first began to find currency in England. The best discussion of these developments is still to be found in Tout, *Chapters in the Administrative History of Medieval England*, iii. 136–9. It is doubtful, incidentally, if Sir William would have needed to obtain the permission of the owners of these coats of arms before displaying them in his church: heraldry was not yet subject to any systematic regime of control or regulation.

royal family on one side and of the earls on the other? He was not one to shrink from asserting either his own importance in the local community or his sole responsibility for the rebuilding of the church. But the gallery of arms sheds light not only on Sir William's personality but also on his understanding of the world in which he lived. For him the ranks of society corresponded to the successive levels within the structure of the parish church itself. The progression from the highest to the lowest, from the sanctuary through the chancel to the nave, was reproduced in the heraldic scheme in the similar progression from the king and the royal family, past the earls to the knights and esquires of Sussex down in the nave. There was an unconscious acceptance that the hierarchy of human society was but a reflection of that in the world to come.

If Sir William saw society in these conventional enough terms, he nevertheless, as we have seen, did not hesitate to remind the onlooker constantly of his own place in it. The arms of Etchingham were to be found in the windows of both the nave and the chancel; they were even incorporated—a nice touch, this—in the weather vane that still rises above the parapet of the tower. Reticence was a characteristic of neither the man nor his church. Indeed, when we look at Etchingham Church, we are reminded of just how far the knightly class had advanced in self-awareness during the previous century or two. In the earliest days of the Norman settlement their fortunes had coincided with those of the tenant-in-chief from whom they held their lands—so much so, that they are found associating themselves in his acts of piety. If he founded a monastery, his vassals would join him in endowing it. It would become the spiritual centre of the entire honor, as Stoke Priory, for example, did for the honor of Clare.[35] Archbishop Theobald's confirmation charter tells us that when Gilbert fitz Richard de Clare founded Stoke-by-Clare Priory, he encouraged his barons to give to the monastery as much as they desired of their lands and churches, provided only that they did not impoverish their successors; and three at least are known to have responded to his call.[36] In Sussex, Battle Abbey, though a royal foundation, may have come to hold a similar position in the affections of the counts of Eu

[35] C. Harper-Bill, 'The Piety of the Anglo-Norman Knightly Class', in *Proceedings of the Battle Conference, II, 1979*, ed. R. A. Brown (Woodbridge, 1980), p. 67; R. Mortimer, 'The Beginnings of the Honour of Clare', in *Proceedings of the Battle Conference, III, 1980*, ed. R. A. Brown (Woodbridge, 1981), p. 140.

[36] J. Ward, 'Fashions in Monastic Endowment', *Jnl. Eccles. Hist.* 32 (1981), 435.

and the honorial baronage of the rape of Hastings. Certainly, Sir William de Etchingham's ancestors are found witnessing grants made to the abbey by the counts and countesses of Eu.[37] Such involvement was a natural expression of the identity of interest that bound tenant-in-chief and vassals. The latter shared with him the administration of the honor, and would have spent as much, perhaps more, time in his company as on their own fiefs. It is only to be expected, then, that they would also share his religious aspirations and would agree to bestow such benefactions as they could afford on the houses which he or his ancestors had founded.

By the thirteenth century, however, this harmony of outlook had begun to dissolve. The widely scattered feudal honor no longer commanded the loyalty from its tenants that it once had. For one thing, the Crown had chosen to make the county rather than the honor the unit through which it governed England. The central and local courts flourished; the honorial courts wilted.[38] Secondly, the knights were spending less time in the company of their lord. It is possible that they felt the need to be seen more on their own estates if they were to retain the respect due to them as landed proprietors. But, more to the point, they needed to learn the arts of estate management. For generations they had been content to live on the fixed incomes provided by the farming, that is to say, the leasing, of their manors.[39] However, the acute inflation of the early thirteenth century compelled them to turn to direct management instead; and direct management entailed constant personal supervision by the lord, constant attention to the details of administration. Now, if for reasons like these the knights were paying more attention to their home base, might they not also have been tempted to take more interest in their local church? The burgeoning of interest in the parish churches might therefore be seen as a by-product of the process by which the gentry came to identify less with the feudal honor and more with their own locality. They were becoming figures of importance in their local communities, and the decision to rebuild a parish church might be taken to indicate that they realized as much. It says something for their perception of their own standing in local society.

[37] Charters of Battle Abbey in the Fuller Collection in the University of London Library (typescript, London, 1979), nos. 2, 24. See also above, p. 141–2.

[38] As a general statement, I think this is true. But in Sussex the existence of the rapes at an intermediate level between shire and hundred complicated the position, and the realignment of loyalties may have proceeded more slowly than elsewhere.

[39] R. Lennard, *Rural England, 1086–1135* (Oxford, 1959), pp. 105–212.

It says something for that perception, too, that in the thirteenth century they began to seek commemoration by the laying of a monumental effigy over the resting-place in the church. Sometimes, as at Bristol, these effigies were placed in recesses along the aisle walls, and sometimes, as at Westminster and Tewkesbury, under the arches surrounding the high altar. Certainly it was burial next to the altar which was most eagerly sought, and by the end of the century major rebuilding programmes were occasioned by the need to satisfy this demand. At Bayham Abbey in the 1260s the church was given an opulent new presbytery to provide more burial space for the Sackvilles.[40] At Robertsbridge a few decades later a new choir was built which, though less spacious than Bayham's—it measured only four bays to the latter's total of seven—bore no less clearly the hallmarks of patronal munificence. It was stone-vaulted and adorned with bosses and carved mouldings of the highest quality.[41] Given its likely date, this work must surely have been paid for by the fourth Sir William de Etchingham, the benefactor whose kindness to the monks we have already had occasion to note.[42] All the same, this new extension, though grander than its predecessor, did not give Robertsbridge anywhere near as much room for burial as as Bayham, and as the number of tombs increased, so too did the likelihood that shortage of space would once more become critical. It is a fair possibility, then, that the rebuilding of Etchingham parish church was undertaken by Sir William V at least in part with a view to providing the additional burial space his family lacked in the abbey. Etchingham was to be

[40] This presbytery is a superb piece of work. Its quality can be savoured from the illustrations in S. E. Rigold, *Bayham Abbey* (London: HMSO, 1974), p. 8. Whether the tie between the Sackvilles and Bayham continued to be as strong in the fourteenth and fifteenth centuries is hard to say. When the claustral ranges were excavated between 1973 and 1976 a number of heraldic titles were uncovered, but none bore the Sackville arms (A. Streeten, *Bayham Abbey* (Sussex Arch. Soc. Monograph, 2, 1983), pp. 75–6). Sir Thomas (d. 1432) expressed a wish to be buried in the abbey (*Testamenta Vetusta*, i. 221–2). On the other hand, his descendants Humphrey (d. 1488) and Richard (d. 1524) opted for the village church at Withyham. (I owe this information to Fr. Jerome Bertram.) This gradual shift of interests from the monastic to the parochial echoes that of the Etchinghams and of the Gloucestershire families noticed above, pp. 142–3.

[41] The church has never been excavated, but its plan can be reconstructed, David Martin tells me, from crop markings visible in aerial photographs. It measured about 210 feet long internally. The architectural fragments are ones that have been recovered from the site.

[42] See above, p. 142, for burials which are known to have been made there. Unfortunately, the burial places of Simon the rector, Sir Robert, and Sir James are not known.

their new necropolis—a family mausoleum no less than a place of worship, its chancel planned on a scale spacious enough to accommodate burials for many generations to come (Plate 2).[43]

It was in that chancel that Sir William himself was buried, in 1389, some fifteen or twenty years after its completion. Over his grave before the altar step was placed a brass showing him in armour under a single canopy (the latter now lost).[44] In design it is a fairly standard product of the London style-B workshop—graceful and elegant, but a trifle monotonous and not, of course, a portrait (Plate 1).[45] However, what the effigy itself may lack in interest, the two inscriptions more than make up for. It would be wrong to claim too much for them, because the brass was not after all laid down until after Sir William had died; but they do contain significant departures from what had become the norm by the late fourteenth century, and in so far as these may have resulted from suggestions made to the engravers by those who had known him, they may afford some insight, however oblique, into the workings of his mind.

Over the head is a semi-circular plate in the shape of a scroll which records that Sir William, the son of Sir James de Etchingham, had been responsible for rebuilding the church 'in honour of God and of the Assumption of the Blessed Mary and of St Nicholas.' Whether these were the saints to whom the church had always been dedicated or were the choice of Sir William himself we cannot tell, but either way they are not entirely without significance as a guide to where his sympathies lay. The other inscription, at the foot of the brass, is in French, and runs to four lines:

De terre fu fet et fourme et en terre fu retourne: William de Echinghm estoie nome, dieu de malme eiez pitee: Et vous qi par ici passez pur

[43] According to *VCH Sussex*, ix. 216, the length of the chancel and the presence of the rather fine stalls together suggest a possible intention of making the church collegiate. Maybe so, but there is no documentary evidence to support the hypothesis. The length of the chancel, if not the quality of the stalls, is better explained in terms of the need to accommodate tombs. It is worth mentioning in this connection that, when Sir William's slab was taken up in the eighteenth century, the stone coffin in which he had been interred was found underneath (W. Slater, 'Echingham Church', *SAC* 9 (1857), 352).

[44] The head of the effigy and presumably, too, the canopy had already gone when Hayley visited the church in 1778.

[45] For the main London workshops see J. P. C. Kent, 'Monumental Brasses—A New Classification of Military Effigies *c*.1360–*c*.1485', *Jnl. Brit. Archaeological Assoc.* 3rd Ser. xiii (1949), 70–96. John Blair identifies Kent's style 'B' in the late fourteenth century with the workshop of Henry Lakenham. See W. J. Blair, 'Henry Lakenham, Marbler of London, and a Tomb Contract of 1376', *Antiquaries Jnl.* lx (1980), 66–74.

lalme de moy pur dieu priez: Qui de Januere le xviii jour de cy passai lan notre seignour mil trois Centz quatre vintz oept come dieu volait entour mynoet.

The opening line replaces the more common formulae like 'Hic iacet . . .' or 'Orate pro anima . . .' It is found in the later fourteenth century on the brasses of at least two London workshops, implying that it enjoyed at least a measure of popularity with clients.[46] Its echoes of the words of the Ash Wednesday liturgy very likely accorded well with the increasing emphasis on austerity, particularly funerary austerity, urged by the more radical thinkers of the day.[47] There is a hint of that feeling of contempt for the flesh which is found in some wills at around this time. The inscription continues in more routine vein by asking the passer-by to pray for the soul of the deceased. The request is, of course, to be found on every medieval inscription; but it would be wrong for that reason to dismiss it as mere convention. Most people had a vivid enough fear of the pains of purgatory to want to seek atonement for their sins by the performance of good works in their lifetime and to ease the passage of the soul after death by arranging for the intercession of the faithful. Some inscriptions went so far as to offer the reader a reward for his prayers, in the case of that on the brass of Etchingham's neighbour, Sir William Fiennes, at Herstmonceux by the granting of 120 days' pardon.[48] Etchingham's executors did not see the need to be quite so generous on his behalf. Doubtless they thought that by rebuilding the church he had already done all that he could to assist the passage of his soul. But prayers did matter. And when considering the reasons that may have led him to undertake the rebuilding, let us not forget that it was concern for salvation that ultimately lay behind all investment in piety.

How essential this great pile was to the religious observances of the founder and his family is, however, another question altogether, and one that it is difficult to answer. For at the heart of the problem lies a paradox: namely, that the evidence we have discussed which docu-

[46] Other brasses on which this wording is found are those of Sir John de Mereworth (*c*.1370) at Mereworth (Kent) and Sir John de Cobham (*c*.1365) at Cobham (Kent). See A. C. Bouquet, *Church Brasses* (London, 1956), p. 157.

[47] For the Ash Wednesday liturgy see *A Catholic Encyclopaedia*, ed. W. E. Addis and T. Arnold, revised by T. B. Scannell and P. E. Hallett (London, 1955), p. 54. I am grateful to Martyn Wakelin for this reference.

[48] The brass at Herstmonceux is described and illustrated by C. E. D. Davidson-Houston, 'Sussex Monumental Brasses, III', *SAC* 78 (1937), 87–8. It commemorates the man who was sheriff of Surrey and Sussex in 1398/9, at the end of Richard II's reign.

ments the strong and increasing commitment of the gentry to the parish church as a family burial place is complemented by evidence (no less compelling) from other sources which attests a shift in the emphasis of gentry religion away from public worship in the parish church towards private worship in the domestic chapel of the manor-house. At the very time that churches like Etchingham were being rebuilt devotion was retreating behind closed doors. It was becoming 'privatized'; or, as Colin Richmond has recently put it, non-communal.[49]

The popularity of domestic chapels sprang partly from convenience, partly from consciousness of status and partly from changes in devotional practice. Convenience is undoubtedly what led to their being built in the first place—the sheer convenience of having a chapel close at hand which relieved proprietors who lived far from the nearest church of the need to attend services there in bad weather.[50] But what had begun as a facility conceded to a few soon developed into a right sought by many. By the fourteenth century permission to build a chapel was being requested by lords of every rank who (if they had cared to be honest) could claim no better justification than that they saw in them a symbol of social status. The traceried window above the altar, the elaborate mass books, the plate, the vestments, the furniture, and doubtless too the ubiquitous coats of arms—all these could be, and were, made to reinforce the owner's sense of social position.[51] Chapels, in short, had become fashionable.

However, it was not all a matter of public display. Quite the contrary. Domestic chapels may also have satisfied the desire for greater privacy that found expression in several areas of upper-class life in the middle ages—for example, in the retreat of the lord and lady

[49] C. Richmond, 'Religion and the Fifteenth-Century English Gentleman', in *The Church, Politics and Patronage in the Fifteenth Century*, ed. B. Dobson (Gloucester, 1984). An article full of valuable insights.

[50] Distance from the nearest parish church was the justification offered in the earliest grants of permission for the use of a private chapel (J. R. H. Moorman, *Church Life in England in the Thirteenth Century* (Cambridge, 1945), p. 15).

[51] They certainly reinforced Thomas of Woodstock's sense of *his* social position, as Jeremy Catto shows in 'Religion and the English Nobility in the Later Fourteenth Century', in *History and Imagination, Essays in Honour of H. R. Trevor-Roper*, ed. H. Lloyd-Jones, V. Pearl, B. Worden (London, 1981). It is worth mentioning, too, that possession of a chapel was taken to be a mark of gentility. When an examination was made touching the gentility of the Paston family, one of the chief pieces of evidence in support of their claim was the fact that since 'time out of mind their ancestors have had licence to have a chaplain and have divine service within (their house)'. H. S. Bennett, *The Pastons and their England* (Cambridge, 1922), p. 226.

from the communal life of the hall to the less public one of the solar or withdrawing room.[52] This process coincided with a change in devotional practice, the consequences of which tended in the same direction. Since the early thirteenth century, if not before, the faithful had been encouraged to seek reconciliation with Christ and his Church through the sacrament of penance known as confession.[53] At first not very many, it seems, did so, but during the thirteenth century, largely as a result of the preaching of the friars, the position improved. The friars, as we know, drew large and enthusiastic audiences; and at the end of their sermons they made a point of rousing their listeners to come forward to confess their sins. Many did so there and then. But many others—and here we are thinking particularly of the gentle-born—preferred to hold back in favour of consulting the preacher afterwards in private. After all, the systematic examination of the conscience was an exercise that could only be carried out in conditions guaranteeing confidentiality between confessor and penitent. It was an exercise moreover which, if it was to be at all effective, would have required not just a single meeting but a continuous process of consultation between the two parties. It is not surprising, then, that it led in so many cases to the appointment of the confessor to a permanent place in the advisee's household.[54]

That appointment need not necessarily have led to the building of a private chapel; but in practice it seems that it usually did. At the very least it implied the existence of a room set aside exclusively for devotional use where the penitent could reflect in relative solitude on his relationship with Almightly God and the saints. The room that lent itself most readily to such use was of course the chapel—assuming that one existed. In the case of the grander houses it can probably be assumed that one did. In the case of their humbler counterparts it is more difficult to say. It was in these latter properties—the dwellings of the gentry—that the imperatives of the new inward-looking religion probably provided the greatest stimulus to chapel-building. Quite a

[52] M. Girouard, *Life in the English Country House: A Social and Architectural History* (New Haven and London, 1978), pp. 40–53.

[53] Moorman, *Church Life in England*, p. 87.

[54] Catto mentions a few examples in 'Religion and the English Nobility in the Later Fourteenth Century', p. 50. These were all mendicants; and I have laid the emphasis on the mendicants in my own account. They were, it seems, highly favoured by the royal family and the titled nobility. But in the households of the gentry they were almost certainly outnumbered by the seculars. The most famous representative of the latter is of course the Pastons' James Gloys.

stimulus it was too.[55] By the end of the fourteenth century there was scarcely a manor-house in England that did not contain a chapel within its walls. And the consequence of that was to relieve the owner and his family of the obligation any longer to attend the local parish church.

Once we realize how peripheral the parish church was becoming to the mainstream practice of late-medieval gentry religion we can begin to understand why the authorities took such care, when licensing the use of private chapels, to insist that the beneficiary should worship in the mother church of the parish at least four times a year.[56] They feared that if they did not so insist the latter would suffer both spiritually and materially. But in the event their fears turned out to be misplaced. Lords continued to go to church: how often we do not know, but continue to go they did.[57] The reason was that, from the social and psychological point of view, their visits, however occasional, had a value that was more than purely religious. They introduced into the ceremonies of the church the social priorities of the world outside. The gentry sat apart—increasingly apart—from their tenants. They had their own pews, and by the fifteenth century their own *family* pews, or 'closetts', screened off from the body of the church.[58] They wanted to be separate from their fellow parishioners; they wanted to suggest to them that the secular hierarchy was, as it were, ordained by God and not made by man. Whether they were aware of the wider implications of what they were doing, and of how their actions might be viewed by their fellow parishioners, is hard to

[55] Discussing the background to the Lollard movement, McFarlane wrote that the late fourteenth century was a period 'much given to private devotions, to private chapels in the houses of the laity, the privilege of appointing one's own confessor with a portable altar and no parochial responsibility'. (*Lancastrian Kings and Lollard Knights* (Oxford, 1972), p. 225). That sentence could serve as the text for much of this and the previous paragraph.

[56] See, for example, *Lewes Cartulary*, i. 132–3. For a grant which was made conditional on attendance at the parish church on at least eleven days in the year see Moorman, *Church Life in England*, p. 15.

[57] The evidence is discussed by Richmond, 'Religion and the English Gentleman', p. 198.

[58] There are two good examples of such 'screened-off' pews in Wootton Wawen church (Warwickshire) one on each side of the arch under the central tower. The Pastons had their own private pew in Paston church. Once again the best discussion of all this is Richmond's in 'Religion and the English Gentleman', p. 198, where mention is made of the dispute between Thomas Rode and William Moreton 'concerning which of them should sit highest in church'—Astbury church (Cheshire), which had a huge parish.

say. Some perhaps were; but many others were probably acting upon impulses that remained inarticulate. All that we can say for certain is that, as an inevitable consequence of the Church's place in the world, attending a service in church became an act of social as well as of spiritual significance. And the same can be said, surely, about the rebuilding of churches. In the case of Etchingham (to return to where we started) the fabric and decoration of the new building were conceived by the patron to bear visual witness to his family's place in the pecking order of local society. It was a building with a message; and that message can still be read today.

VI

THE FAMILIES' LIFE-STYLE

WE may say of Sir William de Etchingham as is said of Wren on the inscription commemorating him in the crypt of St Paul's, 'Si monumentum requiris, circumspice'. His monument, like Wren's, is the church that he built. It is a witness to his ideas and intentions. Yet Sir William's role in the design and construction of *his* building is a less certain one than Wren's in his. Wren was after all the architect, and Sir William only the patron. For all that Etchingham Church bears the undoubted stamp of his personality, from the heraldry on the weather-vane down to the arrangement of the coats of arms in the windows, the respective contributions of patron and master mason cannot be sharply distinguished. The building's witness is a solitary one—solitary, because there are no surviving documents that can help us to place it in its historical context, no accounts that can tell us how much it cost or how long it took to build. As he was paying the bills, Sir William may be presumed to have kept a record of how much was spent. But whether he did so on separate fabric rolls or, as is more likely, on the account rolls on which he recorded his everyday household expenditure we do not know: for all such accounts have gone.

To the extent that they were smaller and simpler, the households of the gentry were also less specialized and less formal than those of the nobility: and perhaps, therefore, less bureaucratic in their approach to business.[1] But there can be little doubt that they did keep weekly accounts of income and expenditure.[2] These were of lowly status certainly—scribbled somewhat in the manner of a modern shopping-list on the back of anything that came to hand, and thrown away as soon as the accounting period was over. Consequently, they are now rarities. Only one has come down to us in the Glynde archives: a

[1] See above, p. 99.

[2] There is a reference to a 'day-roll' in the Glynde account for 1347/8: 'Et in expensis domini et familie sue a dominica predicta ante festum Sancti Edmundi usque festum Sancti Michelis sicut continetur in rotulo dierum £17.12.0.' (GLY 1072, 'Expense hospitii'). These documents are discussed in J. M. Thurgood's introduction to 'The Account of the Great Household of Humphrey, First Duke of Buckingham, for the year 1452–3', ed. M. Harris (*Camden Miscellany*, xxviii, Camden 4th Ser. 29, 1984).

kitchen account of the Waleys household covering the ten weeks from 2 November 1382 to 10 January 1383, written—very neatly it should be added—on the dorse of a notarial instrument of Chichester Cathedral of thirty years earlier.[3] It contains few surprises. It records merely how much bread was baked, barley malted, animals bought and slaughtered, and so on—statistics that are not very interesting in their own right, but which can be used to make a rough estimate of the number of people in the household at the time. Bread, for example, was baked at the rate of six or seven bushels of wheat a week for the first three weeks and a quarter a week for the remaining seven—that is, the weeks immediately before and after Christmas. On the assumption of a daily allowance of the equivalent of 3 lb. per person, this would suggest an establishment of some 15 men and women in November, rising to perhaps 20 as the Christmas festivities drew nearer. These figures are not impossible, but equally they are not foolproof, because they make no allowance for bread given away in alms or, for that matter, fed to the animals. A safer guide is probably afforded by the brewing of ale. For the first four weeks ale was consumed at the rate of 100 gallons a fortnight and thereafter at roughly the same amount a week. Assuming that each person drank the standard 1 gallon a day there must have been about 7 or 8 servants resident in November and double that number in the Christmas period.[4] These figures are perfectly plausible on the assumption that the lord was absent during November and then returned for a lengthy sojourn over Christmas. Indeed, an establishment of a dozen and a half is about what we would expect of a reasonably well-to-do knight.[5] At any rate there is no reason to believe that Sir William Waleys allowed his household to become a source of runaway extravagance.

This document apart, our sources are mainly estate documents—title deeds, rentals and compotus rolls: impersonal perhaps, even a little stereotyped, but nevertheless capable of springing a surprise or two. For instance, the early Beddingham accounts include a number

[3] GLY 4a. The notarial instrument is very faded, but it appears to concern a testamentary endowment of the prebend of Arlington and the maintenance of the fabric of Arlington church.

[4] For the assumptions on which this calculation is based see C. Dyer, 'English Diet in the Later Middle Ages', in *Social Relations and Ideas*, ed. T. H. Aston, P. R. Coss, C. Dyer, and J. Thirsk (Cambridge, 1983), pp. 192–3. I am very grateful to Dr Dyer for the assistance he has given me in interpreting this document.

[5] For the size of noble households see M. Girouard, *Life in the English Country House* (New Haven and London, 1978), p. 15.

of expenses charged for convenience on the issues of that manor which properly belong to a central account—retaining fees paid to lawyers, expenses incurred in going to a funeral and, most interestingly, the termly fees of two boys, John Lucas and William Long, whom Sir William de Etchingham IV sent to school at Lewes.[6] Here are areas of daily concern only hinted at elsewhere.[7] But it is fair to say that the impression given by the mere fact of the survival of all these manorial documents is that estate management itself must have been the chief preoccupation of those for whom they were compiled. Here, on every membrane, is apparently to be found evidence of decisions taken by the lord, of matters reserved for his judgement and of payments made on his authorization—'per talliam domini', 'per litteram domini', and so on in the vocabulary of the accounts. But it is difficult to tell how much of the decision-making as a whole a lord would typically reserve for himself and how much he would delegate to others. It all depended on the size of the estate and on the temperament of its owner. A single manor could have been run by the proprietor and his reeve unaided; half a dozen manors scattered across several counties could not. A steward would have been needed, and probably a full-time stock-keeper as well. So the well-to-do knight was always on the look-out for reliable officials of the stamp of William de Etchingham's John Harold. And it may well have been that William IV's purpose in sending these two lads to school at Lewes was to groom future employees for his service. If a lord could delegate, the chances are that he would. For there were many other, and to his mind better, things to do than worry about wheat prices and flooded ditches. As Mr Keen has recently reminded us, 'tournaments and crusading and far journeys were very much part of the life of [the] nobility, more important to their style than milling or the keeping of sheep . . .'[8]

To the extent that the exercise of arms was both the function of the nobility and the justification for their privileges it is obviously true that tourneying and crusading mattered more than milling or sheep-keeping. But enjoyment of the style of life we associate with Chaucer's

[6] These are mentioned in the Beddingham account rolls for 1307/8 'Expense garcionum'; 1308/9, 'Denarii liberati domino'; 1309/10, 'Expense forinsece' (all in GLY 996).

[7] For example, further light is shed on this question of educational patronage by the tradition preserved in a submission made by the Waleys daughters in the fifteenth century that members of the Waleys family had been 'found to scole at Cambrigg an° xii Edwardi tercii atte costus of the said Sir John (Waleys)' (GLY 25).

[8] M. Keen, *Chivalry* (New Haven and London, 1984), p. 156.

Knight need not necessarily have been incompatible with a display of entrepreneurial flair now and again. In the thirteenth century, for example, the knights were no less eager than any other lords to obtain grants of market rights on their manors. Whether their descendants were to thank them for their initiative is another matter, for the failure rate was high. The market which the Etchinghams had established at Salehurst in 1268 was facing extinction less than a century later.[9] The then head of the family could have accepted its decline as irreversible; but he did not. He and his successors assisted in promoting recovery; and the measures they took provide an object lesson in seigneurial resourcefulness.

It is doubtful if the market at Salehurst had ever amounted to much; but if it did, it was in trouble by the mid-fourteenth century. In 1349 Sir James de Etchingham complained to the king that a dam raised across the Rother at 'Knellesflote' was preventing ships from coming up the river to unload their wares at Salehurst.[10] Perhaps it did; but his neighbours were quick to point out that it also saved their own lands from being flooded.[11] In the long run a more serious problem was the presence of a rival market at Robertsbridge less than a mile away on the opposite bank of the river. This had been founded by the abbot of Robertsbridge on a small spur just off the main north–south route on which Salehurst itself was situated. So successful did the township become that traders preferred to cross the river there rather than at Salehurst further to the east. Eventually this short cut itself became the main route, and Salehurst was left high and dry. Faced with this threat to their income from tolls, the Etchinghams responded by allowing the establishment of a new settlement immediately opposite Robertsbridge at what was to become known as Northbridge. By the mid-fifteenth century its main street was lined with well-built craftsmen's houses.[12] It never became a large place; but it brought a measure of prosperity to the Salehurst traders who settled there, and

[9] D. Martin, 'An Architectural History of Northbridge and Salehurst Villages', in *Historic Buildings in Eastern Sussex*, 2, i (Rape of Hastings Architectural Survey, 1980), p. 1. The income from the fair was down to £1. 10*s*. 0*d*. per annum by 1387 (account of beadle of Etchingham, Hastings Museum JER/Box 3).

[10] *CPR 1348–50*, p. 80.

[11] Ibid. 177–8.

[12] This fascinating story is told by D. Martin, 'An Architectural History of Northbridge and Salehurst Villages'. A sketch-map is given at p. 2. What cannot of course be established in the absence of any documentary evidence is the exact part played by the Etchinghams, as lords of the manor, in the promotion of the new settlement. For that reason I have settled for the neutral word 'allowing'.

The brass of Sir William de Etchingham V
(died 1389) in Etchingham church

Etchingham church showing the stalls and on the floor the brasses of Sir William V and Sir William VI

amply justified whatever investment the Etchinghams might initially have had to make.

The willingness of knights, if not to dabble in trade themselves, then at least to capitalize on the activities of those who did, should hardly occasion any surprise. There was ample opportunity in medieval society for the pursuit of individual style and initiative. The difficulty, if there is one, lies less in recognizing the diversity of interests of the medieval gentry than in trying to measure the relative importance to them of each; of determining how families like ours chose to distribute their time between town and country, between family and society, and between office-holding and the exercise of arms—in a word, to savour something of the general quality of life as they experienced it. Now this is easier said than done, because we do not know that the priorities of the time are necessarily reflected accurately in the incidence of survival of the records. These are, as we have seen, mainly estate documents, and estate administration, though important, may not have been the concern uppermost in every proprietor's mind. But they are all that we have, and we must make the best use of them that we can.

In fact, as we have seen, they can be made to yield quite a harvest of useful information: not all of it individually significant, but nevertheless valuable when taken as a whole. Sadly, it cannot be measured quantitatively, and if used qualitatively it is open to the objection that interpretations based on it are bound to be subjective. Perhaps they are. All history is bound to a greater or lesser degree to be subjective: that is because it is a dialogue between the historian and his subject. But if the dialogue is conducted in the language of imagination, we may be able, as Professor Trevor-Roper has put it, 'to live for a moment as the men of the time lived'.[13] Imagination, in this enquiry at least, may count for as much as scholarship or method—imagination vivid enough to create images of the past, yet sufficiently controlled to prevent them from getting out of focus.[14] These images will not add up, like a series of cartoon frames, to an unfolding narrative. They will be stills. But what they may lack in mobility and drama, they should make up for (it is hoped) in range and depth.

[13] H. R. Trevor-Roper, 'History and Imagination' in *History and Imagination: Essays in Honour of H. R. Trevor-Roper*, ed. H. Lloyd-Jones, V. Pearl, and B. Worden (London, 1981), p. 365.

[14] Ibid. 368. Perhaps it is appropriate to quote Professor Trevor-Roper again: 'For in the end, it is the imagination of the historian, not his scholarship or method (necessary though these are), which will discern the hidden forces of change.'

Our first picture or tableau will be of life at home. For the most part the dwellings in which these families lived are homes no longer. The great castle Sir Edward Dallingridge built at Bodiam still survives, but as an empty shell. Scotney, rebuilt a few years earlier by Sir Roger Ashburnham, lives on as a picturesque fragment in the incomparable setting of Edward Hussey's landscape garden. There are other and older castles, of course—those at Hastings and Pevensey, held usually by grantees of the crown, and at Lewes, held by the Warenne earls of Surrey. But eastern Sussex, for all its exposure to foreign, mainly French, aggression, never became castle country in the way that the northern counties did.[15] It was a landscape characterized not by the mighty fortress but by the unassuming manor-house, girded by a moat fed from a nearby stream. The moat afforded all the protection that was needed by a residence of this kind. But it carried one main disadvantage—the low-lying location dictated by the need for a supply of running water exacted a heavy price in unhealthiness, and for much of the year the occupants were condemned to live in damp, and dank, surroundings. Small wonder, then, that as soon as more peaceful conditions arrived in the sixteenth and seventeenth centuries their descendants removed to higher and drier ground nearby. Sometimes, as at Laughton and Buckhurst, a wing or a gatehouse survives to attest the presence of the older dwelling, but more often, as at Bramblety, there are only earthworks in the grounds of its successor.[16]

A couple of these deserted sites have recently been excavated. They are Hawksden, a property of the Waleys family near Mayfield, and Glottenham, Sir Robert de Etchingham's house near Robertsbridge, both of them secondary seats of their respective families, and each benefiting from discussion in relation to the other.[17] Hawksden's architectural history is more straightforward than Glottenham's. The place was built in the middle of the fourteenth century by Sir John Waleys probably for use as a hunting lodge. It was conceived from the

[15] A list of Northumberland castles compiled in 1415 gives over a hundred names (N. Pevsner, *Northumberland* (Harmondsworth, 1957), p. 42). Sussex does not begin to compare with this. In the fifteenth century Herstmonceux was rebuilt on a grand scale by the war captain Sir Roger Fiennes, but the employment of brick rather than stone indicates that it was conceived as a residence rather than a fortress.

[16] For Bramblety see W. H. Godfrey, 'Brambletye', *SAC* 72 (1931), 1–19. Earthworks somewhat resembling those of a moat are also to be found below the later manor-house at Pashley. They may represent the site of the earlier dwelling there.

[17] I have benefited from being able to read in advance of publication David Martin's report on his excavations of these sites. The report is to appear in *SAC* 125 (1987).

start as a fully developed courtyard house with timber-framed ranges set on all four sides of a paved quadrangle. Access was gained via an entrance passage in the north range. Facing it was the hall, which was probably aisled. In the east range lay the retainers' quarters and next to them the kitchen complex, a set of three rooms separated from the rest of the accommodation by a pair of masonry walls presumably intended to serve as fire-breaks. The entire ground floor of the west range was taken up by the solar and guest lodgings.

In its ground plan Hawksden bore a strong similarity to the castle Sir Edward Dallingridge was to build forty years later at Bodiam—except that at the latter the kitchen was placed, more sensibly, next to the hall.[18] Glottenham's lay-out, on the other hand, was less logical and less symmetrical, because it was the product not of a single building programme but of piecemeal development. The earliest identified structure on the site, dating from the late eleventh or twelfth century, was a long building—a hall perhaps?—to which a cross-wing was later added. A little to the south lay a sizeable detached building measuring 9.5 m by 9.5 m, which is judged to have been a kitchen. If so, it was a remarkably large one for a dwelling occupied by a family—the de Glottenhams—who were of comparatively modest means.[19]

By the 1290s the property had passed into the ownership of Petronilla, daughter of John Andrew, and her husband Robert de Etchingham;[20] and within a few years a building programme had commenced which had the effect of converting it, after a fashion, into a courtyard house. It was a far from straightfoward job. Because the existing buildings could not be demolished until their successors were ready, the masons had to erect the new buildings around and outside the old. They began work by digging a trench 3 m wide in which to lay

[18] I shall not discuss Bodiam in detail because it is already well known. There is a plan in Lord Curzon's volume on this castle, which he saved, restored, and bequeathed to the National Trust (The Marquess Curzon of Kedleston, *Bodiam Castle, Sussex* (London, 1926), opposite p. 52).

[19] One member, Robert de Glottenham (alive in the 1240s), is known to have assumed the rank of knight (*Cal. Robertsbridge Charters*, no. 192). His son Adam, who does not seem to have been knighted, was probably the last of his line to have lived at Glottenham. The family is discussed in more detail in my historical introduction to David Martin's report on the excavation, to appear in *SAC* 125 (1987).

[20] The earliest evidence of their occupation is afforded by a fine levied in 1299 by Robert de Etchingham and his wife Petronilla settling the manors of Glottenham (Sussex) and Holwist (Kent) and other lands in the two counties on the heirs of their bodies with remainder to the right heirs of Petronilla (*Feet of Fines for the County of Sussex from 34 Henry III to 35 Edward I*, ed. L. F. Salzman (SRS vii, 1908), no. 1127). It is clear from the remainder clause that it was Petronilla's property to dispose of.

the foundations of the curtain wall. The soil from this trench was then used to level the interior of the enclosure. Next they started excavating the moat, a massive job entailing the removal of almost 3,500 m^3 of soil which must have taken them as long as the rest of the building works. Whilst all this earth was being shifted, the gatehouse, hall and other domestic quarters were constructed. Then, and only then, were the old buildings demolished and the site cleared.

The new manor-house formed an irregular rectangle 61.00 m long by 38.00 m wide on the east and 45.75 m wide on the west. Entering it from the west the visitor would have crossed a bridge of three bays and passed through a stone-built gatehouse consisting of a block of rooms perhaps two storeys high set one on each side of the archway against the inside of the curtain wall. The buildings within the courtyard included a hall with porch on the north side and a kitchen on the south; those on the east cannot now be identified because that side of the site was so extensively robbed out when the house was dismantled.

Robert de Etchingham was a man of more ample means than the earlier occupants of Glottenham.[21] And the higher standard of living which he and his wife enjoyed is attested by the archaeological evidence. When they first moved in, they did what they could to improve the appearance of the old property by constructing two cinder-paved areas, one in the form of a courtyard in front of the buildings, the other outside the rear doorway of the hall. Moreover, they disposed of their rubbish more tidily than their predecessors had by burying it in pits rather than just discarding it on the ground; and they used high-quality polychrome jugs imported from Saintonge in south-west France. But the house itself—the new house—fell a little short of the highest contemporary standards of grandeur. The outer defences were fine enough. The quoins at the corners of the curtain walls were fashioned from neatly cut stone; and the junction between the substructure and the superstructure was marked by a nicely executed half-roll string-course. But the use of timber-frame construction for the dwellings within might be taken to indicate a concern for economy. Wood was not necessarily a second-choice material in this part of the world. On the contrary, it could be used to give some impressive architectural effects: at Warbleton rectory, probably built at about the same time by a younger son of the lord of Warbleton, the mouldings of the hall truss were as highly finished as those in any

[21] For his career in the service of Edward II and the Despensers see above, pp. 51–5.

cathedral, and the walls, which were of oak boards, were ornamented at least in part with blind arcading.[22] But in a forested area like the Weald timber undoubtedly came cheaper than stone. There was no point in spending money for the sake of it; and Sir Robert himself, though prosperous, was not a man of unlimited means. He was a younger son who had done well for himself, but not that well.[23] He wanted comfort and security, but realized that he would have to forego extravagance. Masonry work, therefore, was confined to the construction of the curtain walls, where it was essential for reasons of pride, and for such practical considerations as defensive strength and resistance to arson. It is doubtful, however, if these walls, however well constructed (and with or without crenellations[24]), could ever have withstood a serious assault. That was not their purpose. Sir Robert was seeking not to defy the siege-craft of an advancing army but to fend off the assaults of lesser enemies nearer home. A man who had spent a lifetime in the service of the Despensers would have had more than enough of them locally: neighbours he had robbed or oppressed, rival landowners whose jealousy he had aroused. These were the people from whom he was seeking protection.

Glottenham is only one of a considerable number of moated sites in this north-eastern corner of Sussex. In the rape of Hastings as a whole no fewer than fifteen have been identified.[25] In the immediate vicinity of Glottenham itself these include the manor-house at Etchingham itself, of course, and the sites at Boarzell, near Ticehurst, and at Great Wigsell a few miles to the east in Salehurst—a remarkable concentration for an area of no more than ten or twelve square miles. Less can be said about these other sites than about the two we have discussed in detail because they have been only surveyed and not excavated. Dates of occupation and abandonment are unknown; and the documentary record is too meagre to be a substitute for the archaeologist's report. At different times in their respective histories, however, Boarzell and

[22] I owe this information to David Martin, who has taken photographs of the features still visible inside the house. (It is now a private dwelling.)

[23] He had done well for himself by marrying a local heiress; but he had not done as well as, say, the elder Sir John Hastings or Sir Thomas Blount, who had picked up heiresses at court (Nigel Saul, 'The Despensers and the Downfall of Edward II', *EHR* xcix (1984), 9–11). It is also worth bearing in mind that life at court could be a source of expense as well as a source of profit.

[24] There is no surviving licence to crenellate—but that does not mean that it was not crenellated, only that no royal authorization was recorded.

[25] This is a high figure, but not as high as in East Anglia, where as many as fifteen are found in two or three parishes. I owe this point to David Martin.

Wigsell, like Glottenham, are known to have been owned by the Etchingham family. The former had figured in a list of manors in which Sir William III had obtained rights of free warren in 1253.[26] The latter was held a little later by one Simon de Etchingham—not the future rector of Herstmonceux, but someone else whose relationship to the other members of the family is difficult to place. In 1348 it was conveyed by a later Simon and his wife to Sir John Culpeper; but oddly enough Simon is still found as a co-parcener in 1372.[27] The descent of these two properties is therefore obscure, to say the least. But that very obscurity is a clue, surely, to their possible significance. They lay in, or were attached to, manors held by the Etchinghams; and so long as they were occupied by members of that family, changes of occupancy or tenancy would have been made orally or by deeds which have not survived. Only if ownership were to pass to another family would the transaction have been registered by fine in Common Pleas. In the case of Wigsell that happened in 1348.[28] Until then the pattern of use implied by the evidence is consistent with what we know of the history of Glottenham after the death of Sir Robert—that is to say, that on the termination of a tenancy the property would revert to the head of the family for him to re-grant to some other kinsman who lacked an establishment of his own. In 1331 that person happened to be James, nephew of Robert's brother and heir Simon, and a young newly-wed. Joan, his bride, was still in the care of her guardian, and James himself could scarcely have grown out of his teens.[29] But, being married, the couple were entitled to a household of their own, and their uncle (in an uncharacteristic fit of generosity) gave them Glottenham in which to establish one. And there, we may presume, they stayed until a few years later James himself succeeded to the inheritance and moved to Etchingham.[30]

The final phase of Glottenham's history is lost to view.[31] But the

[26] *C CH.R 1226—57*, p. 416. The grant was reissued by Edward I when he stayed at Udimore in 1297 (*C Ch.R 1257—1300*, p. 461). This was presumably an act of favour to his host.

[27] *CCR 1307—13*, p. 165; *The Manor of Etchingham cum Salehurst*, ed. Sir S. P. Vivian, 16–17. Vivian is surely right in saying that the first Simon is not the rector: he occurs too early. But neither he nor I can fit him into the family pedigree as we have it.

[28] *Feet of Fines for the County of Sussex, From 1 Edward II to 24 Henry VII*, ed. L. F. Salzman, no. 2076.

[29] Ibid., no. 1788.

[30] James succeeded Simon, the rector, in or after 1337.

[31] It appears in a list of properties which Sir William V settled jointly on himself and his uncle and the heirs male of his body (*Feet of Fines for the County of Sussex, From 1 Edward*

suggestion may at least be ventured that for another generation or two it continued to provide accommodation for junior members of the Etchingham family. William V, like his great-uncle, had no fewer than three younger brothers to think about; and the availability so close to Etchingham itself of these several substantial homesteads would have enabled him to reconcile his instinct to do the best he could for his kinsmen with the obligation to keep the bulk of the inheritance intact for his eldest son and heir.[32]

The physical environment in which our families lived is admittedly hard to visualize. So much has gone, and what remains is often too fragmentary to move any but the most fertile imagination. The chill draught and the damp can be imagined, but probably not the precious objects with which people brightened their surroundings—the wall hangings, coverlets, and plate, for example, that testators singled out for mention in their wills. The surviving wills of members of our three families unfortunately add little to the familiar picture we already have of rooms sparingly but brightly furnished. Sir Thomas Sackville for one was decidedly unhelpful when he said in his will of 1432 simply that he would leave all the goods—unspecified—in his house and wardrobe to his son Edward.[33] Sir William Etchingham V's daughter Joan—the wife of Sir Arnald Savage—was a little more precise: she bequeathed a set of grey furs and a new white gown to Lady Eleanor Cobham and a set of furs called 'Throtes' to one Eleanor Horne.[34] Seven years later her son was principally to be concerned with the salvation of his soul.[35]

A far more valuable insight into the living conditions of the time is to be found in a fascinating inventory of the contents of Sir Andrew Sackville's wardrobe drawn up by his feoffees shortly after his death in

II to 24 Henry VII, ed. L. F. Salzman, no. 2269) in 1362. The final reference to it in Etchingham ownership comes in the year 1415 (PRO, CP40/615). By the mid-fifteenth century it was disused, to judge from the archaeological evidence.

32 Robert, who was sheriff in 1390/1, lived at Great Dixter, for which house see J. E. Ray, 'Dixter, Northiam. A Fifteenth-Century Timber Manor-House', *SAC* 52 (1909), 132–55. The illustrations are valuable for showing Dixter before Lutyens restored it for Nathaniel Lloyd. But where did John and Richard, the other brothers, live?

33 The will is printed in *Testamenta Vetusta*, i. 221–2.

34 G. O. Bellewes, 'The last Savages of Bobbing', *Arch. Cant.* xxix (1911), 167.

35 *The Register of Henry Chichele, Archbishop of Canterbury, 1414–1443*, ii, ed. E. F. Jacob (Oxford, 1938), pp. 205–6.

1370.[36] It forms one paragraph in a lengthy memorandum of the goods and chattels and sums of money which his widow received from the feoffees. A bed with canopy and hanging curtains worth £2, a pair of velvet-lined plates of armour worth 5 marks, several coverlets, a chain with her late husband's seal and two of his horses were among the many chattels that she took from them. The various items were not aranged in any order, and the list of furred robes was entered, almost as an afterthought it seems, on the dorse, between a note about the fattening of sheep at Chalvington and a stock inventory of Newland. It makes more interesting reading than either, but also poses more severe problems of interpretation. First and foremost there is the obvious one of trying to form visual images of the garments being described; and secondly that of evaluating them as evidence of the wearer's standing in society. These difficulties are compounded by the changes in meaning that some of the terms underwent in the fourteenth century. Three of the garments, for example, were called 'gounes': there was a short one of 'scarlet' furred with 'gris', another short one in black and white mi-parti furred with 'gris', and finally a longer one—at least, it is not described as short—furred with 'calabria'.[37] These must have been garments in the fuller, more voluminous style that became fashionable in the 1360s. The word 'goune' itself had been used in a general sense for decades; but in the years after Poitiers it was employed more specifically to describe a tunic which, according to the author of the *Eulogium*, was not open down the front in the proper masculine style but extented at the sides up to the arms so that the wearer, when viewed from behind, could not be distinguished from a woman.[38] Now, writers of old-fashioned inclination have often been heard to complain that it is impossible to tell the sexes apart. Perhaps sometimes it is. But in the case of the 'goune' the *Eulogium* author was probably being less than fair to the tastes of men of his own age and generation. The new style was a good deal more comfortable to wear than the one featuring tight-fitting jupons which had preceded it. Furthermore, it was certainly more flattering to the waist-line of a middle-aged man like Sir Andrew

[36] SAS/CH 258. I wish to thank Kay Staniland for the valuable assistance she has given to me in interpreting this document.

[37] 'Scarlet' at this time describes the quality of the cloth rather than its colour. For 'gris' and 'calabria' see E. M. Veale, *The English Fur Trade in the Later Middle Ages* (Oxford, 1966), pp. 133–41.

[38] The passage is quoted by S. M. Newton, *Fashion in the Age of the Black Prince: A Study of the Years 1340–1365* (Woodbridge, 1980), p. 54.

Sackville. It offered the advantage of combining comfort with the kind of appearance appropriate for a man of his standing.

Over his 'goune' Sackville would probably have worn one of his cloaks. Two are mentioned in the inventory, one of sanguine-coloured cloth furred with 'gris', the other red and black mi-parti furred with pured miniver. The latter would have ranked among the most valuable of the garments in his wardrobe, for miniver—that is to say, the pale winter belly of the red squirrel—was a thicker fur of higher grade than 'gris' or 'calabria'. But better still was to come. The final garment listed was a white surcoat furred with miniver with which went long ermine sleeves or cuffs. Two pairs of such cuffs, in fact, are mentioned—they were probably, in the manner of mid-fourteenth-century fashions, detachable. Their presence in the Sackville wardrobe presents something of a puzzle, because ermine was a fur which royalty usually reserved for themselves. Only seven years before this inventory was drawn up, the sumptuary legislation had prescribed in considerable detail what grades of fur could be worn by each rank of society. It said that knights with incomes of up to 200 marks a year could wear clothing worth 6 marks, but not gowns 'furred with pure miniver or cuffs of ermine'; and that knights with incomes of between 400 marks and £1,000 per annum could wear anything that pleased 'except ermines, letuses and stone apparel, except on their heads'.[39] Ermine, then, was specifically excepted in each category. Yet here was Sir Andrew sporting it on the cuffs which went with his best surcoat. It is clear from the absence of prosecutions under the legislation that he ran no risk of indictment if spotted wearing these cuffs in public. But the question remains: how could he have acquired them in the first place?

The most likely explanation is that they were a gift. As we have seen, Andrew was very much a careerist knight, on the payroll of no fewer than four magnates, and an active participant in the opening stages of the Hundred Years War. Now one of the rewards of service in the middle ages was the gift of a furred robe usually bearing the badge of the lord who bestowed it. Some of the cloaks in his wardrobe might have been such robes of livery, and the cuffs might have come with them.

Of one thing at least we may be sure. Attired in this clothing he would have cut a very impressive figure indeed. The nearest visual

[39] *Statutes of the Realm*, i (1810), 381.

approximation to his appearance is probably to be found in the costume worn by the two unknown civilians on the brass at King's Sombourne (Hants.) and by the priest's brother at Shottesbrooke (Berks.)—remembering always that the tarnished latten surface scarcely does justice to the vivid colours of the original cloths.[40] These garments, of course, would have been the very best in Sir Andrew's wardrobe, and it is likely that for everyday wear he would have donned something older and less fashionable which would never have been mentioned in an inventory. The garments we have been discussing were the ones he would have worn when summoned to attend his lord: the 'goune' by itself indoors, a cloak bearing his employer's livery almost certainly thrown over it outdoors when he was travelling on official business.

For Sir Andrew was someone who spent a good deal of his life out on horseback. He was a busy man, busier than most of his neighbours. His itinerary took him to Scotland on active service in 1336, to the Low Countries in the same cause in 1338, back to England, to Colnbrook (Bucks.), by September 1339, then to Shropshire on the earl of Arundel's business early in the following year and finally back to Flanders a few months later.[41] Few could match that level of activity. But most could at least claim to have crossed the Channel a few times, and some to have gone a good deal further. In the 1370s Andrew Waleys went on a pilgrimage to Jerusalem from which he was never to return;[42] and in 1380 his brother William obtained a licence to travel to the papal court at Rome.[43] These were the 'far journeys' that Mr Keen tells us mattered more to the gentry than milling or the keeping of sheep. How many more such journeys they undertook we cannot tell, not at least from our Sussex sources alone. But the submissions to the Scrope–Grosvenor hearings in the 1380s show how widely travelled members of the northern gentry were. William, a younger son of Chief Justice Sir Geoffrey Scrope, had fought in the earl of Hereford's retinue at Satalia on the Turkish coast, and his nephew, another Geoffrey, had died on crusade in Prussia in 1362. Stephen, who was ultimately to succeed as lord of Masham, served in France, and after

[40] The brass at King's Sombourne is illustrated in H. W. Macklin, *The Brasses of England* (London, 1907), p. 59.

[41] PRO, E101/19/36 m. 1, C76/12 m. 6, C76/14 m. 9, C76/15 m. 20; *CPR 1338–40*, pp. 348, 395, *1340–3*, p. 336.

[42] GLY 19.

[43] *CPR 1377–81*, p. 546.

1360 followed in his brother's crusading footsteps. He was present at the taking of Alexandria, where he was knighted by King Peter of Cyprus; and later he was to serve in Prussia.[44]

In the careers of these knights reality merged with the world of heroic literature. The men who went to fight in Prussia, Professor Du Boulay has written, were at home almost anywhere. They acquired their tastes at an early age, and the slight discomforts and occasional dangers they suffered were as nothing against the comradeship and luxuries of the officers' mess.[45] For them the crusade was a living reality not a relic of the past. It might take the form of smaller expeditions than before, each with a more specific objective. But the appeal it exerted and the response it evoked remained as potent as ever, apparently little diminished by the passing years.

Clearly the knights of northern England who gave evidence in the Scrope–Grosvenor hearings stood in the very forefront of the crusading movement. Schooled as they were in the hardships of border warfare, soldiering was a way of life to them. For that very reason, of course, their own involvement in crusading may well have been much greater than that of the gentry resident in the more peaceful counties of central and southern England. But if it was—and it is a big 'if'—that is not to say that the latter led lives which were either sedentary or uneventful. On the contrary, they lived a restless existence spent constantly on the move. Alongside Sir Andrew Sackville's itinerary for the late 1330s we may set Sir William de Etchingham's for 1309/10 as glimpsed—but no more than glimpsed—through the Beddingham account roll for that year.[46] Sir William passed through, or more likely stayed overnight in, the village almost once a month: in January and March, in late April while on his way to a meeting of the county court in Chichester, in June on the way to and on the way back from his wife's manor of Stopham, again in July and August, and finally in September at harvest-time for a full week. Here we see the manors of an estate becoming staging-posts in the regular annual perambulations of the household. Inasmuch as this journeying was

[44] These examples are taken from M. H. Keen, 'Chaucer's Knight, the English Aristocracy and the Crusade', in *English Court Culture in the Late Middle Ages*, ed. V. J. Scattergood and J. W. Sherborne (London, 1983), pp. 51–2.

[45] F. R. H. Du Boulay, 'Henry of Derby's Expeditions to Prussia, 1390–1 and 1392', in *The Reign of Richard II*, ed. F. R. H. Du Boulay and C. M. Barron (London, 1971), p. 155.

[46] The account roll is in GLY 996. The details are taken from the paragraph marked 'Expense domini'.

occasioned by the needs of social intercourse and estate management, its parameters were defined largely by the extent of the family's own lands—that is to say, eastern Sussex and Kent in the case of a family like the Etchinghams, further afield in the case of one with more scattered possessions like the Sackvilles. But periodically the needs of business or politics caused the itinerary to be extended. In a typical year in the early fourteenth century, for example, Sir William de Etchingham would have needed to fit in journeys to Scotland on active service with the king, to Westminster for a session of parliament, and to Horsham and Lewes, as well as to Chichester, for meetings of the county court. Horsham was a good thirty miles from his home, and Chichester at least fifty.

Lewes, on the other hand, was on his doorstep. But if it was, from his point of view, the most convenient of the three venues, it was probably in this period the least assuming. Whatever importance it had it owed largely to the presence in its midst of the powerful Warenne family, and when that presence was removed, as it was in 1347, its life-blood drained away.[47] There was still the priory, of course: its numbers temporarily depleted after the Black Death, but its hold on the affections of the local community probably little weakened.[48] As a focus for the activities of the local gentry, however, it could not compare with a magnate household. It is true that it was a source of employment for ambitious young officials—like John Harold, perhaps?—and that it remained in the fourteenth century as popular as it had ever been as a place of burial.[49] But its presence was not sufficient to tempt members of the local gentry to acquire property in the town. At least it was not sufficient to tempt the Etchinghams—or the fifth Sir William would not have had to impose himself, as he did in 1378, on the hospitality of his steward John Harold, who owned a house in the suburb of Cliffe.[50] The Etchinghams may not have been

[47] In the 1380s the castle was used as a gaol. It fell to an assault by a gang of malefactors protesting against the exaction of labour services (*CPR 1381–5*, p. 259).

[48] The number of monks fell slightly from 33 in 1306 to 32 in 1350, but then rose again to 58 in 1391. On the eve of the Dissolution, however, there were only 24 (*VCH Sussex*, ii. 66, 68).

[49] See above, pp. 42.

[50] GLY 1000: 'Et in probend' equi domini ad domum Johannis Harold in Blyva 2 bus avene. Et in probend' equi domini existentis ibidem ut patet per talliam 6 bus.' John Harold was a native of Etchingham. His decision to move to Cliffe, while at the same time remaining in the service of the Etchinghams, must have been occasioned by the offer of employment there. Who could the offer have come from? The prior of Lewes is certainly one possibility. But it could equally have been the archbishop of Canterbury,

arbiters of fashion; but if *they* saw little point in owning property in Lewes, their neighbours would surely have seen even less.

Legally, in fact, there was little to be gained from owning property there. It is true that the town was a parliamentary borough, and that it enjoyed a measure of internal self-government.[51] But its status as a manor of the Warennes would probably have limited the privileges extended to the holders of tenements. This is a point worth stressing because country gentlemen did not buy town houses merely to economize on overnight accommodation expenses. They had friends with whom they could stay; they had manors of their own in the vicinity from which they could travel in.[52] Nor did they buy such houses in order to acquire a base for shopping expeditions. The servants after all could be dispatched on errands to the local market-places—as they were by the Etchinghams in 1307/8 to Shoreham to buy some fish, and in the following year to Chichester to acquire clothes for the lord and to Seaford to pick up uniforms for the two lads sent to school in Lewes.[53] If a gentleman bought a town house, it was for more than personal convenience: it was for prestige, perhaps for legal and political gain. In medieval Sussex such interests could best be served by moving into Winchelsea.

The old-world buildings of present-day Winchelsea are in fact those of a medieval new town. By the 1280s old Winchelsea lay in such severe danger of inundation by the sea that Edward I made plans to transfer it to a hill-top site at Iham, about three miles further up the Brede river.[54] Fresh from their achievements in town-planning in Gascony and North Wales Edward and his advisers knew exactly how

who was lord of the manor of Cliffe (across the river from Lewes) in which John chose to live.

[51] There is a solitary reference to the existence of a merchant guild at Lewes (*Lewes Cartulary*, ii. 25). Royal mandates to Lewes were addressed sometimes to the bailiffs and burgesses but more usually to the bailiffs and good men, a formula first used in 1266. Thus although the burghal status of Lewes before and after the Conquest is unmistakable, according to *VCH Sussex*, vii. 24, the late middle ages provide no further evidence as to its constitutional development.

[52] On one occasion in 1387/8 Sir William de Etchingham stayed at Beddingham and rode in to Lewes for a session of the assizes: 'Et in expensis domini existentis apud Bedyngham ad assisam apud Lewis, 17s 8d.' (GLY 1002, 'Expense domini').

[53] GLY 996, account rolls for 1307/8, 'Expense garcionum' and 1308/9, 'Denarii liberati domino' and 'Expense nunciorum et forinsece'.

[54] For a lively account of the story told in the context of Edward's other new towns see T. F. Tout, *The Collected Papers of Thomas Frederick Tout* (3 vols., Manchester, 1932–4), iii. 59–91.

to go about their task. They laid out the site generously in thirty-nine blocks or 'quarters' divided by streets intersecting at right-angles, and each 'baron', or freeman, of the town appears to have been assigned one plot at the harbour, one in the northern part of the town for a residential and business site, and one at the southern end perhaps for manufacturing or storage purposes.[55] A few of the new residents, however, contented themselves with taking just a single tenement in the residential area. These were the local gentlemen—Sir Roger de Lewknor, Sir William de Etchingham and his kinsman Sir Simon, all next door to each other in the quarter adjacent to the Grey Friars.[56] They did not envisage importing or manufacturing: that much we can deduce from their failure to take up tenements on the quayside and at the southern end of town. But as buyers and sellers in the market-place they still had much to gain from establishing a foothold in the new settlement. They had a right to trade in the new town, to arrest debtors on sight and to extract compensation for damage inflicted on their own property by citizens of another town.[57] These were privileges enjoyed by any freeman of the borough. But it is also fair to say that they were the kind of privileges enjoyed by the freemen of most other boroughs. What made Winchelsea different, what made it more receptive than other neighbouring towns to outsiders, was that it was starting afresh.[58] The very fact of uprooting made it easier for outsiders to make their contribution to the collective identity of the new town. And once established there, they stayed put.[59] At Rye, by contrast, there were no resident country gentlemen in the early

[55] W. M. Homan, 'Winchelsea: the Founding of a Thirteenth-Century Town' (1940, typescript in ESRO).

[56] A rental compiled in 1292 lists the holders of the original tenements. It is printed in W. D. Cooper, *The History of Winchelsea* (London and Hastings, 1850), pp. 44–52. Homan used this rental to reconstruct the lay-out of the plots in the town. Copies of his maps can be consulted in ESRO. The quarter where the Etchinghams lived is now deserted. Their neighbour Sir Roger de Lewknor held the manors of Horsted Keynes and Selmeston (Sussex), Greatworth (Northants.), and South Mimms (Middlesex), and was probably a retainer of the Clares (Moor, *Knights*, iii. 31–2).

[57] These were rights worth having: we must not forget that Winchelsea was an important centre for the disposal of demesne produce from local manors such as Udimore. See for example the Udimore account roll for 1380/1 (Hastings Museum, JER/Box 3).

[58] It is also worth mentioning in this connection that Winchelsea was the only royal borough in this part of Sussex—a point made to me by Christopher Whittick.

[59] The Etchingham tenement was still in the hands of Sir William's descendants in 1414 (RYE 146/2).

fifteenth century.[60] Winchelsea, then, may have been the more important of the two towns not only economically but also socially and politically. If it was not a provincial capital, it was more than an ordinary port or market town. It was where the local urban and rural élites came together.

For come together they certainly did—to discuss the prevention of flooding, the hiring of ships, the defence of the coastline, the payment of tolls and a host of other matters of mutual concern.[61] Even before it became common, as it did in the fifteenth century, for country gentlemen to represent boroughs in parliament, the local squirearchy and the burgesses felt sufficiently at ease with each other for acquaintance to ripen into friendship—and for friendship between their respective offspring sometimes to ripen into romance. The young Robert de Etchingham, we know, went so far as to repudiate a pre-contract with a county gentleman's daughter in order to marry a Winchelsea girl by the name of Petronilla Andrew. Her father, to be sure, was a wealthy man who had a foothold in both camps. A freeman of Winchelsea by origin, he had invested some of his money in the purchase of a number of small estates in the surrounding countryside, one of which—Glottenham—lay so close to the manor of Etchingham that it may have been there that the two young people first developed an affection for each other.[62] In espousing Petronilla, therefore, Robert was neither marrying beneath himself nor moving outside his own social circle. But he was insisting on the right to choose his own partner, even to the extent of breaking an earlier engagement. His action may not have been to the initial liking of his elder brother, who was soon to find himself sued for breach of contract by the ex-fiancée's mother and stepfather.[63] But in the long run the Etchinghams stood to gain: Petronilla, being an heiress, was a far better catch than the daughter of a minor—very minor—country squire.

[60] A scot and half-scot imposed in about 1414–15 provide useful lists of residents at Rye at that time (RYE 77/1 and 2).

[61] These are simply a few examples taken from *CPR 1307–13*, p. 601; *1324–7*, pp. 302, 310; *1338–40*, p. 362; *1343–5*, p. 515; and *1339–1401*, p. 346.

[62] See above, p. 166–9.

[63] The record entered on the roll of Common Pleas for Hilary term 1306 is our main source for this episode (PRO, CP40/158 m. 188d). John de Burne and his wife Christine, respectively the stepfather and mother of Margery, Robert's ex-fiancée brought an action for debt against Sir William de Etchingham IV alleging that he owed them £100, being the sum due to them consequent on the breaking of the engagement. Sir William said that he had already paid it. The jurors' verdict was in his favour.

Gentry marriage alliances were usually contracted with local families, be they of urban or gentle origin. For the Etchinghams Winchelsea counted as a local town, perhaps the local town. What fits less easily into this conventional picture of relationships is the marriage arranged by the respective parents between John Waleys II of Glynde and Joan, daughter of the London fishmonger Sir Robert Turk.[64] Through her mother Joan had inherited country estates in Hertfordshire and Hampshire; but the family were Londoners first and foremost. Robert owned property in the capital, and his friends were there—men like Richard Jeep, rector of All Saints, Honey Lane, and Bartholomew Seman, the goldbeater, who both came down to Glynde in June 1398 to witness the delivery of seisin of some manors to the young couple.[65] It was in Sussex, of course, that the latter chose to settle.[66] But they took care to keep up their links with London society. Joan was to find husbands there for a couple of her daughters—William Melreth, a mercer and sometime MP, for Beatrice, and a man by the name of John Burgh for Agnes. In view of her own background she evidently did not experience the hesitations that most gentle-born mothers did about the danger of surrendering a daughter to the insecurity of life without land.[67]

So much is understandable. But how had the Waleyses and the Turks made each other's acquaintance in the first place? We simply do not know. All we can say is that matches between London folk and knights from far afield were by no means as uncommon as we might suppose. Sir Thomas FitzNichol, for example, a contemporary of John Waleys from Gloucestershire, married the widow of a London skinner, and Sir Nicholas Stukeley of Cambridgeshire, a draper's widow.[68] Although prejudice tended to limit intermarriage between gentlewomen and merchants, it did little, as Sylvia Thrupp pointed

[64] They were married in or before 1398 (GLY 5, 6).

[65] GLY 1139, fo. 33, no. 102.

[66] John had been eager to claim some of the family's manors for himself as soon as he had got married. In June 1398 he had received Glynde and Patching from his father and three years later those of Hawksden and Bainden too—by which time he was in possession of the best part of the inheritance. Father and son were clearly at odds for some time, and in 1406 they finally agreed to submit their differences to arbitration. Sir William recovered Glynde and Patching, and his son was confirmed in possession of Hawksden and Bainden. It was in one or other of those manors that he and his wife were residing, presumably, when they succeeded to the inheritance in *c.*1409 (GLY 5, 6, 1140 (v)).

[67] Sylvia Thrupp, *The Merchant Class of Medieval London* (Chicago, 1948), p. 263.

[68] Ibid. 267.

out, to check the marriage of gentlemen with merchants' daughters and widows: the gentlemen were too eager to pocket the fat dowries their wives would bring.[69]

We need not imagine that the medieval gentry were tottering on the edge of bankruptcy to accept that from time to time they needed credit to overcome temporary cash-flow difficulties. The number of outright spendthrifts was small, very small. And at this distance of time spendthrifts cannot easily be distinguished from plain bad managers of money. Into which category, for example, is Sir Roger de Bavent II of Wiston (Sussex) to be placed? The origins of his indebtedness is not clear. But his rake's progress, once begun, proved difficult to halt. On 23 February 1343 in return for a loan of £60 he demised to Robert de Burton, a canon of Chichester, for seven years his manor of Hatcham in Camberwell and his house in Bassishaw St, London.[70] Two weeks later for money lent he demised to two London merchants for ten years his manor of Shipbourne (Kent) with rents in Malling and Rochester.[71] On 9 May he granted his manor of Poling to the earl of Arundel.[72] And finally on 1 July 1344 he surrendered all but two of his manors to the king.[73]

Roger's story is instructive, but not typical. Few found that their needs could not be met by the raising of a single, or at least occasional, loan—to cover, for example, the cost of providing a daughter with a suitable cash portion, of fitting out a contingent for active service or of paying relief on succession to an inheritance. Sources of money were not hard to find. Most knights would have begun, as Sir Roger did, by looking locally. Their own friends and neighbours might be possible lenders. In thirteenth-century Kent members of local gentry families came to the assistance of Christ Church, Canterbury, some of them with quite considerable loans. Henry de Sandwich, for example, lent £495. 6*s*. 8*d*. over a period of thirty years and John de Gatesden £460. 13*s*. 4*d*.[74] Lenders and borrowers, one suspects, were often the

[69] Ibid. 265.

[70] *CCR 1343–6*, p. 95. For the Bavent family see W. Hudson's 'Miscellaneous Notes' to P. S. Godman, 'On a Series of Rolls of the Manor of Wiston', *SAC* 54 (1911), 146–9.

[71] *CCR 1343–6*, pp. 88–9.

[72] Ibid. 112.

[73] Ibid. 521. William de Keynes, who had been appointed keeper of Roger's surviving lands, was ordered to deliver to him the issue of those lands in aid of his maintenance. But Roger's problems were not over. In 1349 he acknowledged in King's Bench that he owed Sir William de Thorp the sum of £60 (PRO, KB27/357 m. 15).

[74] M. Mate, 'The Indebtedness of Canterbury Cathedral Priory, 1215–95', *Ec.HR* 2nd Ser. xxvi (1973), 290.

same people at different stages in their careers. A century later in Sussex we find that the famously rich earl of Arundel, 'Copped Hat', was lending money to some of his neighbours. According to an inventory drawn up after his death in 1376, the abbot of Battle owed him 150 marks, the abbot of Durford 90 marks, the prior of Hardham 100s and Sir Edward St John 60 marks.[75]

But the search for credit could, and did, take men further afield. In the first half of the century the Italian banking houses were sometimes approached. In February 1340 John de Farningham, a knight of Kent, acknowledged that he owed Boniface de Peruzzi and his fellows of Florence the sum of 200 marks.[76] In the wake of the downfall of the Italian firms, however, it was English financiers who came to dominate the market.[77] In Chaucer's day the merchant Gilbert Mayfield was lending considerable sums to the king and to members of the nobility and gentry, among them Sir Edward Dallingridge. His activities are known to us because he later went bankrupt, and his account book passed to the Exchequer.[78] But, as the recognizances on the Close Rolls testify, many Londoners before him had been involved in the operation of credit. Among their clients were several members of our own families, notably Sir Andrew Sackville III. The first debt acknowledged by this knight, in March 1336, was in the sum of £100.[79] It may have been a loan to cover the expenses he had incurred in equipping himself with the accoutrements of knighthood—his career in arms had begun in 1334, when he had attended a couple of tournaments, and in both the following two years he had been to fight in Scotland.[80] But this can only be a surmise: it cannot be proved. The

[75] BL Harl. MS 4840 fo. 395. And, of course, thirty years earlier this same earl had acquired the manor of Poling from an indebted Roger de Bavent.

[76] *CCR 1339–41*, p. 442.

[77] The Peruzzi went bankrupt in 1343 and the Bardi in 1346 (M. McKisack, *The Fourteenth Century* (Oxford, 1959), p. 223).

[78] M. K. James, 'A London Merchant of the Fourteenth Century', *Ec.HR*, 2nd Ser. viii (1955–6), 372. His name, usually spelt 'Maghfeld' in the documents, suggests that he came from Mayfield (Sussex). He lent Dallingridge the sum of £20. For the circumstances see C. M. Barron, 'Richard II and London, 1392–7', in *The Reign of Richard II*, ed. Du Boulay and Barron, pp. 184–5.

[79] *CCR 1333–7*, p. 654.

[80] Andrew Ayton tells me that Sir Andrew Sackville's arms are to be found in the second Dunstable Roll and the Carlisle Roll, both of 1334. For a description of these two rolls see A. R. Wagner, *A Catalogue of English Medieval Rolls of Arms* (Harl. Soc., 100, 1948), 54–7. In 1335 he appointed attornies for the duration of his active service in Scotland (PRO, C71/15 m. 34). And in 1336 he served, again presumably in Scotland, in the retinue of John of Eltham, the king's brother (PRO, E101/19/36 m. 1).

problem with using recognizances is that they are open to so many possible interpretations. What are we to make, for example, of the several bonds into which the same Sir Andrew entered between 1347 and 1349? On 4 June 1347 he and two other knights, Sir Edward de Kendal and Sir Roger le Warde, and a clerk, William de Osberton, acknowledged that they owed Simon Dolseley and John Nott, citizens and pepperers of London, the sum of £800, and three days later the same men, with John de Alveton taking the place of Kendal, acknowledged that they owed another pepperer, John Hammond, the sum of £600.[81] In the following year Sackville entered into two recognizances with a London fishmonger by the name of Richard Double: in the first he and Osberton acknowledged a debt of £600 and in the second he alone acknowledged one of £300.[82] Then a year later in 1349 he, Kendal, Warde, Alveton and Osberton came together again to acknowledge a final debt to Double of 800 marks.[83] The likely explanation for these recognitions of what were very large sums of money is probably to be found in the next set of recognizances to be enrolled. On 9 March 1349 William Montagu, the new earl of Salisbury, acknowledged that he owed that same group of men—Sackville, Kendal, Warde, Alveton, and Osberton—the sum of £400.[84] And six days later Elizabeth Despenser acknowledged that she too owed them the same sum.[85] Now Elizabeth Despenser was William Montagu's sister.[86] And Sackville was a retainer of both families.[87] Roger Warde and Edward Kendal were probably retainers of the Despensers too: they had certainly fought on the Crecy and Calais campaign in the retinue of Elizabeth's late husband, Hugh.[88] In the light of this knowledge it seems reasonable to suppose that the three knights with Alveton and Osberton were acting not as principals but as agents in the raising of loans for their lords. The pepperers, unlike the fishmongers, are not known to have been regular lenders of large sums of money, but they no less than their colleagues had cash to

[81] *CCR 1346–9*, pp. 281, 283. Kendal came from Hertfordshire and Warde from Leicestershire (ibid.). John Alveton was an Oxfordshire gentleman whose daughter married the influential local knight Sir Gilbert Wace (J. E. Titterton, 'The Malyns Family: Medieval Manorial Lords', dissertation submitted for the Certificate in Local History, Oxford, 1985). For Dolseley, Hammond, and Nott, see Thrupp, *Merchant Class of Medieval London*, pp. 337, 348, 358 respectively.

[82] *CCR 1346–9*, pp. 507, 524.

[83] *CCR 1349–54*, p. 57.

[84] Ibid. 58.

[85] Ibid. 61.

[86] GEC, iv. 273.

[87] For a discussion of his career in service see above, pp. 48–51.

[88] *Crecy and Calais*, ed. Wrottesley, pp. 85, 92.

spare. Hammond and his fellow trader Aubrey imported merchandise worth as much as £2,000 a year.[89] Spices were in high demand in aristocratic households, but not surely, to the value represented in these recognizances. They must relate to debts, not purchases.

So far, then, we may go in our speculation, but no further. Recognizances do not yield up their secrets that easily. But of this, if nothing else, we may be certain, that the more active country gentry were no strangers to the London scene. They would frequently have been drawn there on business, whether to see the likes of Dolseley and Nott to raise a loan,[90] of Henry de Lakenham, near St Paul's, to order a tomb slab,[91] or of the mercer Richard Whittington to purchase some fine cloth for the ladies of the family.[92] Some might even have ventured down the Strand to consult a lawyer acquaintance at the Temple—some, but not in the fourteenth century all that many. Consultancy was not yet as common as it was to become 100 years later when a standard rate of 3*s*. 4*d*. a session was to be charged.[93] In the preceding century or two landowners could probably have found most of the legal advice they needed locally. There were attorneys they could approach, like Sir William de Etchingham's Michael de Pyecombe.[94] There were professional pleaders sprung from local families of gentry stock, like Sir Edmund de Pashley in the 1310s and Sir William Tauk in the 1360s.[95] And there were professional pleaders turned country landowners, like William Halden, the recorder of London, who acquired an estate in Sussex and came to be identified with the county closely enough to be appointed to its commission of the peace.[96]

[89] G. A. Williams, *Medieval London: From Commune to Capital* (London, 1963), p. 134.

[90] See above, p. 183.

[91] I mention Lakenham by way of example merely because he has been the subject of a recent and most illuminating article by John Blair, 'Henry Lakenham, Marbler of London, and a Tomb Contract of 1376', *Antiquaries Jnl.* lx (1980), 66–74. The area close to St Paul's cemetery was later to be closely associated with the monumental trades.

[92] For 'Dick' Whittington see C. M. Barron, 'Richard Whittington: the Man Behind the Myth', in *Studies in London History Presented to P. E. Jones* (London, 1969), pp. 197–248. Another reason for visiting London would have been to pay an instalment of a rent. When Sir Philip St Clair of Bramblety took a lease on the manors of Brambletye, Laverty, Jevington, and Heighton (Sussex) from his mother in 1389, the arrangement was that the rent should be paid annually in St Paul's Cathedral at a place called the 'Resurrection' (*CCR 1389–92*, p. 76).

[93] E. W. Ives, *The Common Lawyers of Pre-Reformation England. Thomas Kebell: a Case Study* (Cambridge, 1983), p. 299.

[94] For whom see above, pp. 92–3.

[95] For the Pashley family see N. H. MacMichael, 'The Descent of the Manor of Evegate in Smeeth', *Arch. Cant.* lxxiv (1960), 1–27; and for William Tauk see J. B. Post, 'The Tauke Family in the Fourteenth and Fifteenth Centuries'.

[96] He was Recorder 1365–76 (*Calendar of Letter Books of the City of London*, ed. R. R.

The availability of a pool of legal talent locally would have saved the gentry a good deal of time queueing up in the waiting-rooms of the capital's lawyers, and for that they would have been grateful: for time, if not of the essence, was not to be squandered. The journey from home would have taken a day or more, and as many nights would then have had to be spent in London as were necessary for dealing with the business in hand. For those with a house in the capital accommodation was no problem; but they were very much the minority. Only those compelled to spend several months of the year in London would have considered it worth their while actually to invest in property there. In 1319 Sir Edmund de Pashley, the Exchequer baron, bought a tenement in Lambeth Hill, in the parish of St Mary Magdalen, Old Fish St; but his son, who had not followed in his father's judicial footsteps, made so little use of it that he provided in his will for it to be sold.[97] Sir William de Etchingham IV, a periodic visitor to Westminster if not to London, had a house in the parish of St Katherine, Aldgate; but this, too, was sold off in the next generation by a kinsman who saw no point in its retention.[98] Sir Andrew Sackville, oddly enough, seems to have had nowhere. For a man who, as we have seen, was a frequent visitor to the capital and who chose to be buried in a fashionable City church—Christ Church, Grey Friars—this is a little surprising.[99] But it has to be remembered that his life centred not so much on the court, not so much on London and Westminster, as on

Sharpe (11 vols., London, 1889–1912), passim). He was appointed a JP in Sussex in March 1361 (*CPR 1361–4*, p. 63). He was a feoffee of Sir Robert de Pashley (*CCR 1364–8*, p. 289), and a witness to Roger Dallingridge's acquisition of Sheffield (ibid. 404). He must have lived, or held property, in one of the Etchingham manors of Etchingham, Beddingham, Peakdean, or Udimore, since he is one of the tenants whose homage and services are mentioned in a settlement of those manors in 1362 (*Feet of Fines for the County of Sussex from 1 Edward II to 24 Henry VII*, ed. L. F. Salzman, no. 2269).

[97] Edmund de Pashley acquired the tenement in 1319 from one Thomas de Bottele (Corporation of London Record Office, Index to Hustings Deeds, Roll 48, no. 8). The later Edmund de Pashley provided in his will, dated 1 Aug. 1361, that all his tenements in London, near Old Fish St, were to be sold for payment of debts and for pious uses (*Calendar of Wills proved and enrolled in the Court of Husting, London, A.D. 1258–A.D. 1688*, ed. R. R. Sharpe (2 vols., London, 1889), ii. 59).

[98] Corp. of London Record Office, Index of Hustings Deeds, Roll 71, no. 17; PRO, E210/D.9594.

[99] For knowledge of his place of burial (his will does not survive) we are indebted to Stow (John Stow, *A Survey of London*, ed. C. L. Kingsford (Oxford, 1908, reprinted 1971), i. 332). For the observation on the popularity of the church I am indebted to Caroline Barron.

the households of his magnate employers.[100] The 'country' gentlemen who acquired property in the capital were, by contrast, mainly courtier knights like Sir Robert Sapy, Sir Walter Mauny, and Sir Nigel Loring who spent much if not most of the year (the campaigning season apart) in and around Westminster.[101] Doubtless they had chambers in the palace; but they needed surroundings more spacious, more commodious, for entertaining and receiving guests.[102] So they invested in a house in town: after all, it could be let while they were away.

Whatever the reasons that may have led them to invest in property, these courtier knights constitute the odd exceptions needed to prove the general rule: that most country gentlemen—the 'mere gentry' as they might be called in a later century—did not have a place of their own in the capital.[103] When drawn there on business, they would have had to stay with friends (as one imagines Sir James de Etchingham probably did with the Pashleys until they sold their own house) or to find a room at an inn. The number of inns in London was certainly on the increase in the late middle ages, and in the fifteenth century the Pastons were to be regular patrons of 'The George' on Paul's Wharf.[104] Whether visitors from Sussex had their own regular haunt we do not know; but if they did, it may well have been the 'Herteshed' in Southwark, which is known to have been owned in the early 1400s by a Sussex family, that of Tregoz of Dedisham.[105]

The English gentry, we may therefore say, were visitors to, rather than residents in, the towns. To that extent it is doubtful if they ever penetrated urban society in the way that their Italian or Flemish counterparts did. True, they sometimes received the honour of

[100] My assertion that Sackville did not own a London tenement is based on my failure to find any transactions involving him among the Court of Hustings deeds. It is very difficult to prove a negative; but the silence of the records in this case is surely telling.

[101] Respectively Corp. of London Record Office, Index to Hustings Deeds, Roll 50 no. 6, Roll 79 no. 48, Roll 80 no. 72. Other purchasers of property drawn from the same background were Sir John Beauchamp, Sir John Neville, and Sir Reginald Cobham.

[102] For the earl of Nottingham dining with John Northampton in London see *The Westminster Chronicle, 1381–1394*, ed. L. C. Hector and B. F. Harvey (Oxford, 1982), p. 62.

[103] I am thinking of the 'mere gentry' who figured in the arguments of a generation ago about the 'rise of the gentry' and the origins of the Civil War. From Sussex Sir Roger de Bavent and Sir John Covert were the only other knights whom I have found owning property in London: and Bavent was forced to sell (*CCR 1343–6*, p. 95; Corp. of London Record Office, Index to Hustings Deeds, Roll 75, no. 134).

[104] *Paston Letters and Papers of the Fifteenth Century*, ed. N. Davis (2 vols., Oxford, 1971, 1976), i, nos. 286, 288, 290, 300, 316, 363, 365–8, 375, 404.

[105] L. F. Salzman, 'Tregoz', *SAC* 93 (1955), 55.

election to an urban guild; and true, they were sometimes asked by the townsmen to represent them in parliament. But whether these developments, which were anyway only in their infancy in the fourteenth century, amounted to what Sylvia Thrupp called 'cultural interpenetration' is open to question. The penetration was too much one way. The merchant class contributed little that was uniquely their own; rather, they preferred to assume the manners and values of the rural nobility. They longed for the accolade of knighthood; they adopted heraldic coats of arms.[106] And the wealthiest of their number rounded off their careers by moving into the country and building splendid manor-houses, as Sir John Pulteney did at Penshurst and Laurence de Ludlow at Stokesay. Noblemen were expected to live in fine houses, crenellated to have the air of a castle, and to keep hawks and hounds. Living nobly, said Poggio scornfully, meant wasting your time in the open air, going hawking and hunting.[107] He may not have approved; but he was almost certainly right. Hawking and hunting—particularly hunting—take us to the very heart of the gentleman's life-style.

The popularity of the hunt is amply attested by the literature it inspired in the late middle ages. From *La Chace dou Cerf* to the *Boke of St Alban's* a succession of treatises described how the sport strengthened both morally and physically those who participated in it. Inasmuch as it developed a capacity for endurance, it was esteemed as a means of keeping the body fit and of practising those skills of horsemanship that were essential to the exercise of arms.[108] By the fourteenth century the sport described in these manuals was developing in a number of directions. At its crudest it might involve little more than the digging of a pit towards which the game would be driven and then killed with a knife. As practised by the nobility, however, it had become a more elaborately organized affair, in which ladies participated as well as men.[109] According to one set of conventions,

[106] It is an interesting sidelight on contemporary perceptions that from roughly the middle of the fourteenth century county sheriffs found it worthwhile to cite the names of burgesses in their returns to writs of distraint for knighthood. In 1356 the sheriff of Gloucester listed Walder de Frampton, John de Wycombe and Richard Hurel, all of Bristol (PRO, E159/133 membranes unnumbered). Michael Powicke cites examples from the 1340s of Londoners being distrained and citing tenure by burgage as grounds for exemption (*Military Obligation in Medieval England* (Oxford, 1962), pp. 177–8).

[107] Keen, *Chivalry*, p. 154.

[108] M. Thiebaux, 'The Medieval Chase', *Speculum*, 42 (1967), 260–74.

[109] L. F. Salzman, *English Life in the Middle Ages* (Oxford, 1926), p. 100.

described by the author of *Sir Gawain and the Green Knight*, the beasts were driven down from the hilltops and shot by the noble hunters, who positioned themselves at bends in the glades. Those that survived were chased by the hounds into an enclosure where they were then slaughtered.[110] Less hidebound by rules, but deemed by the writers to be the noblest of all forms of the sport, was the formal chase in which the huntsmen rode at the head of their packs and either dispatched the beast from the saddle or cornered it for the hounds to drag down and finish off. If we bear in mind how dangerous it was to release an arrow from the back of a fast-moving horse, it becomes easier to understand why prowess at hunting came to be regarded by contemporaries as a mark of nobility.[111]

The 'sport of kings' was therefore both an expression of the contemporary social order and a means of supporting it. It not only epitomized the values of a rural élite but provided a regime of training whereby those values could be maintained. It was utilitarian and practical as well as enjoyable; but it was surely the enjoyment which was the source of its enduring appeal. Something of the atmosphere of excitement and expectation that was generated on the morning of a hunt can still be felt across the centuries in the opening of the third fitt of *Sir Gawain and the Green Knight*. Dawn had not yet broken, but already people in the castle were stirring:

> Guests who had to go gave orders to their grooms,
> Who busied themselves briskly with the beasts, saddling,
> Trimming their tackle and tying on their luggage.
> Arrayed for riding in the richest style,
> Guests leaped on their mounts lightly, laid hold of their bridles,
> And each rider rode out on his chosen way.
> The beloved lord of the land was not the last up,
> Being arrayed for riding with his retinue in force.
> He ate a sop hastily when he had heard mass,
> And hurried with horn to the hunting field;
> Before the sun's first rays fell on the earth,
> On their high steeds were he and his knights.
> Then these cunning hunters came to couple their hounds,
> Cast open the kennel doors and called them out,
> And blew on their bugles three bold notes.
> The hounds broke out barking, baying fiercely,

[110] *Sir Gawain and the Green Knight*, lines 1150–77.
[111] F. Barlow, *William Rufus* (London, 1983), p. 123.

And when they went chasing, they were whipped back.
There were a hundred choice huntsmen there, whose fame
Resounds.
To their stations keepers strode;
Huntsmen unleashed hounds:
The forest overflowed
With the strident bugle sounds.[112] (lines 1126–49)

The landscape in which the huntsmen sought their quarry was eerie and dramatic:

> on both sides banks, beetling and steep,
> And great crooked crags, cruelly jagged:
> The bristling barbs of rock seemed to brush the sky. (lines 2165–7)

Were it not for the fact that the poet is known to have come from the north-west, we might be forgiven for thinking that it was Ashdown Forest that he was describing. The steep-sided banks and 'bristling barbs of rock' are as evocative of this Sussex landscape as they are of any remoter part of England. Only the mention of 'great crooked crags' suggests that it was in fact somewhere more rugged that the writer had in mind. Ashdown, though certainly wild and unspoilt, is characterized by unfolding ridges rather than jagged outcrops: it is emphatically a southern and not a northern landscape. For that reason, then, if for no other, it would still be advisable (following Ralph Elliott) to regard Swythamley Park, in the Staffordshire moorlands, as the location with the strongest claim to have been the setting of Sir Gawain's last trial.[113]

It is not without a wider significance, moreover, that Swythamley is a park and not a forest: for in the thirteenth and fourteenth centuries deer parks became for the nobility and gentry what the forests had long been for the king—exclusive preserves where they alone could hunt. Their chief physical characteristics were a substantial earthen bank topped by a pale of cleft oak stakes and an inside ditch usually filled with water.[114] Even though most parks lay on the edge of the manor, well away from the open fields, the scale of earth-moving that

[112] In order to make the poem accessible to a modern audience I have used the translation by Brian Stone in *Sir Gawain and the Green Knight* (2nd edn., Harmondsworth, 1974). A good edition of the original is by R. A. Waldron (York Medieval Texts: London, 1970).

[113] R. W. V. Elliott, 'Staffordshire and Cheshire Landscapes in *Sir Gawain and the Green Knight*', *North Staffordshire Jnl. of Field Studies*, 17 (1979 for 1977), 20–49.

[114] *The English Medieval Landscape*, ed. L. Cantor (London, 1982), pp. 73–80.

was involved and the violation that was likely to be done to the property rights of neighbouring tenants meant that the creation of such an enclosure was an operation that could not be undertaken without a good deal of planning. This Sir John Waleys realized when he embarked on creating the new park at Hawksden. In 1337 he made an agreement with Robert de Sharnden, whose tenants adjoined the proposed enclosure, laying down the procedures that were to be followed in deciding the line of the new boundaries.[115] This was clearly to be no ordinary park. When the boundaries were agreed, and the pales were in place, Sir John went on to equip it with a grand, new hunting-lodge—the one the lay-out and design of which we have already discussed. To judge from its location and date of construction, it could have been intended for use only as a hunting-lodge. But in ground-plan and conception, it amounted to a full-scale manor-house in all but name.[116]

At roughly the same time that he was undertaking these works at Hawksden, Sir John was also carrying out improvements to the park at Glynde. According to the account roll for that manor, he spent the modest sum of £2. 2*s*. 3*d*. in the year 1347/8 on the construction of two wooden lodges to add to what today might be termed its amenity value.[117] The park at Glynde, unlike its counterpart at Hawksden, was a long-established one—very likely as old as the manor-house to which it was attached.[118] For that reason capital spending there was usually very low—in the late 1340s just a couple of pounds on those lodges (which were probably for the benefit of spectators) and that was all. It was principally current spending that had to be met—the parker's stipend, the mowers' wages, the cost of maintaining the perimeter pale and so on.[119] The task of maintaining the pale, in fact,

[115] AMS 5896/5 m. 2, a transcript made at a court held on 23 July 1566 of an agreement between Waleys and Sharnden dated 2 Feb. 1337. I am grateful to Christopher Whittick for drawing my attention to this document. For Sharnden, who was the archbishop of Canterbury's steward, and therefore not someone to be trifled with, see above, pp. 45, 48.

[116] See above, pp. 166–7.

[117] GLY 1072, Glynde account roll 1347/8, 'Custus novi logg''. The main item is a payment to one Richard Machon 'pro 2 logg' in parco de novo fac''.

[118] The earliest surviving reference to Glynde park seems to be in a quitclaim of 29 Apr. 1291: the tenant of a field called 'Breglond' was required annually at harvest-time to provide a man, bearing a white rod, to superintend for one day the reapers in Sir Richard Waleys' park (GLY 1143). Perhaps it is also worth mentioning here that the Etchinghams had at least two parks—one at Etchingham itself, said to be of 400 acres, and the other at Udimore, said to be of 200 acres (PRO, C134/100/15). Both figures, coming as they do from an inquisition *post mortem*, are likely to be underestimates.

[119] At Glynde in 1368 the parker's stipend was said to be 6*s*. 8*d*.; he was also paid a

was a work of higher priority than might be supposed, because most proprietors faced a double problem of not only keeping the deer in but also keeping the poachers out.[120] Parks were a constant source of temptation to trespassers, and the gentry hunted those of the nobility as freely as the lower orders did their own. Sir Andrew Sackville and his son and, of course, Sir John Waleys were among the many offenders in the middle years of the century. In 1352, when Queen Philippa sent her justices of oyer and terminer into the rape of Pevensey, each of these men was indicted several times for hunting illegally in her forest of Ashdown.[121] They paid their fines, set variously at four or five marks, and doubtless continued to hunt exactly as before. Then in 1384, as we have seen, John of Gaunt attempted a clampdown of his own. This time the chief offenders were Sir Edward Dallingridge, his son in law Sir Thomas Sackville and Sir Philip Medsted. The charges included not only illegal hunting but also the murder of a sub-forester. The defendants were found guilty but refused to pay the fines; and the duke could do nothing to enforce their collection. As an absentee landowner there was little he *could* do. Indeed, his very powerlessness is indicated by the fact that during part of the time these trespasses were being committed his master forester in Ashdown was none other than Dallingridge himself: the poacher turned gamekeeper.[122] He and his like knew perfectly well that they could hunt the local parks at will: and they did.

If we were to choose a setting and an occasion with which to draw this study to a close we could do no better than to opt for these hunting parties and the opportunities they afforded for social intercourse. What passed between the participants in conversation our sources never reveal. But we cannot suppose that when men as important locally in their different ways as the prior of Lewes and Sir John St Clair met the earl of Oxford to go hunting, as they did at the earl's

'vadium' of 10½*d*. a week which came that year to £2. 5*s*. 6*d*. (GLY 1073, 'Vadia', 'Stipendia').

[120] It is worth recalling in this context that in 1381 one of the demands of the peasant rebels was that all warrens 'tam in aquis quam in parco et boscis' should be open to all (*Chronicon Henrici Knighton*, ed. J. R. Lumby (Rolls Series, 2 vols., 1889, 1895), i. 137).

[121] PRO, JUST 1/941A mm. 4d, 25, 38d, 45, 45d. Every lord of Ashdown (as of every other forest) had this problem. Three-quarters of a century before Eleanor of Provence, the Queen Mother, had appointed Sir William de Etchingham III and Sir Stephen de Penchester to investigate trespasses that had been alleged to have taken place in the forest at that time (*CPR 1281–92*, p. 105).

[122] See above, p. 75.

manor of Laughton in 1375, they did not take the opportunity to discuss matters of mutual concern.[123] They need not always have talked high politics: though hunting itself could sometimes have been a political act. When Dallingridge, Sackville, and Medsted trespassed in Ashdown, it was not only for the pleasure of the chase; it was to make a gesture of defiance against the lordship of 'a powerful but alien newcomer' to the county.[124] But more commonly the political significance of hunting was covert rather than overt. And with that in mind we might conclude by permitting ourselves this final speculation. If tournaments could sometimes serve to mask baronial gatherings (if not to disguise actual conspiracies), as they did in the thirteenth and fourteenth centuries,[125] could not hunting parties have served a similar function in county society? The social history of hunting has yet to be written; but it will make an interesting read when it is.

[123] BL Add. Roll 32141. The prior at this time was John de Caroloco, who was to be captured by a French raiding party at Rottingdean two years later (see above, p. 43). Sir John St Clair passed from the service of Queen Philippa to that of John of Gaunt when the latter inherited the Queen's estates in the rape of Pevensey (R. Somerville, *History of the Duchy of Lancaster* (London, 1953), i. 380). He died in 1387 holding in Sussex the manors of Brambletye, his chief seat, Laverty, Jevington, Heighton St Clair and Exceat (*CCR 1385–9*, p. 237; W. Bugden, 'Excete and its Parish Church', *SAC* 58 (1916), 144–5).

[124] The phrase is that of Simon Walker, 'Lancaster *v*. Dallingridge', 88.

[125] J. R. Mandicott, *Thomas of Lancaster*, pp. 99–100.

BIBLIOGRAPHY

A. MANUSCRIPT SOURCES

1. East Sussex Record Office, Lewes

GLY	Archives of Glynde Place, including account rolls of the manors of Glynde and Beddingham.
SAS	MS Collections of the Sussex Archaeological Society, including account rolls of the manors of Chalvington, Claverham, and Heighton St Clair.
RYE	MSS of the corporation of Rye.
AMS	Additional MSS, including account roll of the manor of West Dean.

2. Hastings Museum

JER/Box 3	Ray collection, including account rolls of the manor of Udimore

3. Public Record Office, London

C47	Chancery Miscellanea
C67	Patent Rolls Supplementary
C71	Scotch Rolls
C76	Treaty Rolls
C81	Chancery Warrants
C134	Inquisitions *post mortem*
E43	Exchequer, Ancient Deeds
E101	Exchequer, Accounts Various
E179	Subsidy Rolls
CP40	Common Pleas
JUST 1	Assize Rolls
KB27	King's Bench
SC8	Special Collections, Ancient Petitions

4. British Library, London

Add. MS 28550	Fragment of a chronicle of Robertsbridge Abbey.
Add. MS 39373–5	MS notebooks of Edwin Dunkin: notes from the public records.
Add. MS 39490	MS note books of Edwin Dunkin: topographical notes.
Add. Roll 56359	Account roll of the manor of Goring.
Stowe 553	Liber Cotidianus, 1 May 1322–19 Oct. 1323.

5. Corporation of London Record Office, Guildhall, London, Index of Hustings Deeds

6. Lambeth Palace

ED 695	Account of the serjeant of Mayfield Rectory, 1394–5
ED 927	Account of the chamberlain of Ringmer, 1387–8

B. PRINTED SOURCES

Book of Bartholomew Bolney, ed. M. Clough (SRS, lxiii, 1964).

Calendar of Charters and Documents relating to the Abbey of Robertsbridge ... Preserved at Penshurst among the Muniments of Lord de Lisle and Dudley (London, 1873).

Calendar of Close Rolls.

Calendar of Fine Rolls.

Calender of Inquisitions Miscellaneous.

Calendar of Inquisitions Post Mortem.

Calendar of Patent Rolls.

Calendar of Plea and Memoranda Rolls ... of the City of London, A.D. 1323–1364, ed. A. H. Thomas (Cambridge, 1926).

Cartulary of the High Church of Chichester, ed. W. D. Peckham (SRS, xlvi, 1942–3).

Charters of Battle Abbey in the Fuller Collection in the University of London Library (typescript, London, 1979).

Chartulary of the Priory of St Pancras, Lewes (SRS xxxviii and xl, 1932 and 1934).

Chronicon Henrici Knighton, ed. J. R. Lumby (Rolls Series, 2 vols., 1889, 1895).

Chronicon Monasterii Sancti Albani Thomae Walsingham, Historia Anglicana, ed. H. T. Riley (Rolls Series, 2 vols., 1863, 1864).

Cokayne, G. E., *The Complete Peerage*, ed. V. Gibbs and others (12 vols. in 13, London, 1910–59).

Crecy and Calais, ed. G. Wrottesley (London, 1898).

Curia Regis Rolls.

Expeditions to Prussia and the Holy Land made by Henry, earl of Derby, ed. L. Toulmin Smith (Camden Soc. NS lii, 1894).

Feet of Fines for the County of Sussex, from 34 Henry III to 35 Edward I, ed. L. F. Salzman (SRS vii, 1908).

Feet of Fines for the County of Sussex, from 1 Edward II to 24 Henry VII, ed. L. F. Salzman (SRS xxiii, 1916).

Feudal Aids.

Glynde Place Archives, A Catalogue, ed. R. F. Dell (Lewes: East Sussex County Council, 1964).

'Kent Fines, Edward II', ed. J. Greenstreet, *Arch. Cant.* xiii (1980).

Manor of Etchingham cum Salehurst, ed. Sir S. P. Vivian (SRS liii. 1953).
Registrum Roberti de Winchelsey Cantuariensis Archiepiscopi A.D. 1294–1313, i, ed. R. Graham (Cant. and York Soc., 51, 1952).
Register of Henry Chichele, Archbishop of Canterbury, 1414–1443, ii, ed. E. F. Jacob (Oxford, 1938).
Register of John Pecham, Archbishop of Canterbury, 1279–1292, ed. F. N. Davis and others unnamed (Cant. and York Soc. 64, 1969).
Scotland in 1298, ed. H. Gough (Paisley, 1888).
Stow, J., *A Survey of London*, ed. C. L. Kingsford (Oxford, 1908, reprinted 1971).
Taxatio Ecclesiastica Anglie et Walliae, auctoritate P. Nicholai IV, circa A.D. 1291, ed. T. Astle, S. Ayscough, and J. Caley (Record Commission, 1802).
Testamenta Vetusta, ed. N. H. Nicolas (2 vols., London, 1826).
Three Earliest Subsidies for the County of Sussex, ed. W. Hudson (SRS x, 1910).
Walter of Henley and other Treaties on Estate Management and Accounting, ed. D. Oschinsky (Oxford, 1971).
War of Saint-Sardos (1323–1325), ed. P. Chaplais (Camden 3rd Ser. lxxxvii, 1954).

C. SECONDARY SOURCES

Baker, J., *An Introduction to English Legal History* (London, 2nd edn., 1979).
Ball, W. E., 'The Stained Glass Windows of Nettlestead Church', *Arch. Cant.* xxviii (1909).
Barron, C. M., 'The Quarrel of Richard II with London 1392–7', in *The Reign of Richard II*, ed. F. R. H. Du Boulay and C. M. Barron (London, 1971).
— 'Richard Whittington: the Man behind the Myth', in *Studies in London History presented to Philip Edmund Jones* (London, 1969).
Bean, J. M. W., *The Decline of English Feudalism, 1215–1540* (Manchester, 1968).
Belcher, W. D., *Kentish Brasses* (2 vols., London, 1888).
Bellamy, J. G., *Crime and Public Order in England in the Later Middle Ages* (London, 1973).
Bellewes, G. O., 'The last Savages of Bobbing', *Arch. Cant.* xxix (1911).
Bennett, M. J., *Community, Class and Careerism: Cheshire and Lancashire Society in the Age of Sir Gawain and the Green Knight* (Cambridge, 1983).
Blair, W. J., 'Henry Lakenham, Marbler of London, and a Tomb Contract of 1376', *Antiquaries Jnl.* lx (1980).
Bolton, J. L., *The Medieval English Economy, 1150–1500* (London, 1980).
Bouquet, A. C., *Church Brasses* (London, 1956).
Brandon, P. F., 'Agriculture and the Effects of Floods and Weather at Barnhorne, Sussex, during the Late Middle Ages', *SAC* 109 (1971).
Bridbury, A. R., 'The Black Death', *Ec.HR*, 2nd Ser. xxiv (1973).
Britnell, R. H., '*Advantagium Mercatoris*: A Custom in Medieval English Trade', *Nottingham Medieval Studies*, xxiv (1980).

— 'Production for the Market on a Small Fourteenth Century Estate', *Ec.HR*, 2nd Ser. xix (1966).
Budgen, W., 'Excete and its Parish Church', *SAC* 58 (1916).
Cantor, L. editor, *The Medieval English Landscape* (London, 1982).
Carpenter, C., 'The Beauchamp Affinity: A Study of Bastard Feudalism at Work', *EHR* xcv (1980).
Catto, J., 'Religion and the English Nobility in the Later Fourteenth Century', in *History and Imagination: Essays in Honour of H. R. Trevor-Roper*, ed. H. Lloyd-Jones, V. Pearl, B. Worden (London, 1981).
Clanchy, M. T., *From Memory to Written Record* (London, 1979).
Cooke, A. H., *The Early History of Mapledurham* (Oxford, 1925).
Cooper, W. D., *The History of Winchelsea* (London and Hastings, 1850).
— 'Notices of Winchelsea in and after the Fifteenth Century', SAC 8 (1856).
Copinger, W. A., *The Manors of Suffolk* (7 vols., Manchester, 1905–11).
Curzon, the Marquess, of Kedleston, *Bodiam Castle, Sussex* (London, 1926).
Davidson-Houston, C. E. D., 'Sussex Monumental Brasses, I–IV', *SAC* 76–80 (1935–9).
Davies, R. R., *Lordship and Society in the March of Wales, 1282—1400* (Oxford, 1978).
Denholm-Young, N., *Collected Papers in Medieval Subjects* (Oxford, 1946).
Drew, J. S., 'Manorial Accounts of St Swithun's Priory, Winchester', *EHR* lxii (1947), reprinted in *Essays in Economic History* ii, ed. E. M. Carus-Wilson (London, 1962).
Du Boulay, F. R. H., 'Henry of Derby's Expeditions to Prussia, 1390–1 and 1392', in *The Reign of Richard II*, ed. F. R. H. Du Boulay and C. M. Barron (London, 1971).
— *The Lordship of Canterbury: An Essay on Medieval Society* (London, 1966).
Dyer, C., 'English Diet in the Later Middle Ages', in *Social Relations and Ideas: Essays in Honour of R. H. Hilton*, ed. T. H. Aston, P. R. Coss, C. Dyer, J. Thirsk (Cambridge, 1983).
— *Lords and Peasants in a Changing Society: The Estates of the Bishopric of Worcester, 680—1540* (Cambridge, 1980).
— *Warwickshire Farming, 1349—c.1520, Preparations for Agricultural Revolution* (Dugdale Soc., Occasional Paper, 27, 1981).
Emden, A. B. ed., *A Biographical Dictionary of the University of Oxford to A.D. 1500* (3 vols., Oxford, 1957–9).
Fairbank, F. R., 'The last Earl of Warenne and Surrey', *Yorks. Arch. Jnl.* xix (1907).
Fryde, E. B., 'The Deposits of Hugh Despenser the Younger with Italian Bankers', *Ec.HR* 2nd Ser. (1951).
Fryde, N., 'A medieval robber-baron: Sir John Molyns of Stoke Poges, Buckinghamshire', in *Medieval Legal Records, edited in memory of C. A. F. Meekings* (London: HMSO, 1978).

Girouard, M., *Life in the English Country House: A Social and Architectural History* (New Haven and London, 1978).
Godfrey, W. H., 'Brambletye', *SAC* 72 (1931).
Goodman, A., *The Loyal Conspiracy: The Lords Appellant under Richard II* (London, 1971).
Hall, S., *Echyngham of Echyngham* (1850).
Harding, A., *The Law Courts of Medieval England* (London, 1973).
Harriss, G. L., *King, Parliament and Public Finance in Medieval England to 1369* (Oxford, 1975).
Harvey, B., *Westminster Abbey and its Estates in the Middle Ages* (Oxford, 1977).
Harvey, P. D. A., ed., *Manorial Records of Cuxham, Oxfordshire, circa 1200–1359* (Oxfordshire Record Society, 50, 1974, published 1976).
— *A Medieval Oxfordshire Village: Cuxham, 1240–1400* (Oxford, 1965).
Hilton, R. H., *Bond Men Made Free* (London, 1973).
Holmes, G., *The Estates of the Higher Nobility in Fourteenth Century England* (Cambridge, 1957).
Horsfield, *The History, Antiquities and Topography of the County of Sussex* (2 vols., London, 1835).
Hudson, W., 'On a Series of Rolls of the Manor of Wiston', *SAC* 53 (1910).
Hutchins, J., *The History and Antiquities of the County of Dorset* (4 vols., Westminster, 1861).
Ives, E. W., *The Common Lawyers of Pre-Reformation England. Thomas Kebell: A Case-Study* (Cambridge, 1983).
James, M. K., 'A London Merchant of the Fourteenth Century', *Ec.HR* 2nd Ser. viii (1955–6).
Jessup, R. F., *Sussex* (The Little Guides: London, 1949).
Jones, W. R., 'Patronage and Administration in the King's Free Chapels in Medieval England', *Jnl. British Studies*, ix (1969).
Keen, M. H., 'Chaucer's Knight, the English Aristocracy and the Crusade', in *English Court Culture in the Later Middle Ages* (London, 1983).
— *Chivalry* (New Haven and London, 1984).
Knowles, D., *The Monastic Order in England, 940–1216* (Cambridge, 2nd edn., 1963).
Kosminsky, E. A., *Studies in the Agrarian History of England in the Thirteenth Century* (Oxford, 1956).
Lambarde, F., 'Coats of Arms in Sussex Churches, I', *SAC* 67 (1926).
Lloyd, T. H., *The Movement of Wool Prices in Medieval England* (Economic History Review Supplements, 6, 1973).
Lower, M. A.,, *Historical and Genealogical Notices of the Pelham Family* (privately printed, 1873).
Macmichael, N. H., 'The Descent of the Manor of Evegate in Smeeth', *Arch. Cant.* lxxiv (1960).

Maddicott, J. R., *Law and Lordship: Royal Justices as Retainers in Thirteenth- and Fourteenth-Century England* (Past and Present Supplement, 4, 1978).
— 'The County Community and the Making of Public Opinion in Fourteenth-Century England', *TRHS* 5th Ser. 28 (1978).
— *Thomas of Lancaster, 1307–1322* (Oxford, 1970).
Margary, I. D., 'Roman Roads from Pevensey', *SAC* 80 (1939).
Mason, E., 'The Role of the English Parishioner, 1100–1500', *Jnl. Eccles. Hist.* xxvii (1976).
Mate, M., 'The Indebtedness of Canterbury Cathedral Priory, 1125–1295', *Ec.HR* 2nd Ser. xxvi (1973).
McFarlane, K. B., *The Nobility of Later Medieval England* (Oxford, 1973).
McKisack, M., *The Fourteenth Century* (Oxford, 1959).
Miller, E., and Hatcher, J., *Medieval England: Rural Society and Economic Change, 1086–1348* (London, 1978).
Moor, C., *Knights of Edward I* (Harleian Soc., 80–4, 1929–32).
Moorman, J. R. H., *Church Life in England in the Thirteenth Century* (Cambridge, 1945).
Morant, P., *History and Antiquities of the County of Essex* (2 vols., London, 1763–8).
Mortimer, R., 'The Beginnings of the Honor of Claire', in *Proceedings of the Battle Conference, 1980*, ed. R. A. Brown (Woodbridge, 1981).
Nairn, I., and Pevsner, N., *Sussex* (Harmondsworth, 1965).
Neilson, N., 'The Court of Common Pleas', in *The English Government at Work, 1327–1336*, iii, ed. J. F. Willard, W. A. Morris and W. H. Dunham (Cambridge, Mass., 1950).
Newton, S. M., *Fashion in the Age of the Black Prince: A Study of the Years 1340–1365* (Woodbridge, 1980).
Palmer, R. C., *The County Courts of Medieval England, 1150–1350* (Princeton, 1982).
— *The Whilton Dispute, 1264–1380* (Princeton, 1984).
Pelham, R. A., 'The Foreign Trade of Sussex, 1300–1350', *SAC* 70 (1929).
— 'Studies in the Historical Geography of Medieval Sussex', *SAC* 72 (1931).
— 'The Exportation of Wool from Sussex in the late Thirteenth Century', *SAC* 74 (1933).
Perceval, C. S., 'Remarks on some Charters and other Documents relating to the Abbey of Robertsbridge, in the County of Sussex', *Archaeologia*, xlv (1879).
Pevsner, N., *North-West and South Norfolk* (Harmondsworth, 1962).
— *Worcestershire* (Harmondsworth, 1968).
Phillips, C. J., *A History of the Sackville Family* (2 vols., London, 1929).
Phillips, J. R. S., *Aymer de Valence, Earl of Pembroke, 1307–1324* (Oxford, 1972).
Post, J. B., 'The Tauke Family in the Fourteenth and Fifteenth Centuries', *SAC* 111 (1973).

Postan, M. M., ed., *Cambridge Economic History of Europe, I: The Agrarian Life of the Middle Ages* (2nd edn., Cambridge, 1966).
— *The Medieval Economy and Society* (London, 1972).
Powell, E., 'Arbitration and the Law in England in the Late Middle Ages', *TRHS* 5th Ser. 33 (1983).
Powicke, F. M., *The Thirteenth Century* (2nd edn., Oxford, 1962).
'Proofs of Age of Sussex Families', *SAC* 15 (1863).
PRO, *List of Sheriffs* (London, 1898).
Ray, J. E., 'Dixter, Northiam: a Fifteenth Century Timber Manor House', *SAC* 52 (1909).
Richmond, C., *John Hopton, A Fifteenth Century Suffolk Gentleman* (Cambridge, 1981).
— 'Religion and the Fifteenth-Century English Gentleman', in *The Church, Politics and Patronage in the Fifteenth Century*, ed. B. Dobson (Gloucester, 1984).
Rigold, S. E., *Bayham Abbey* (London: HMSO, 1974).
Ritchie, N., 'Labour Conditions in Essex in the Reign of Richard II', in *Essays in Economic History*, ii, ed. E. M. Carus-Wilson (London, 1962).
Roskell, J. S., *The Commons in the Parliament of 1422* (Manchester, 1954).
Round, J. H., 'The Lords Poynings and St John', *SAC* 62 (1921).
Rowney, I., 'Arbitration in Gentry Disputes in the Later Middle Ages', *History*, 67 (1982).
Salzman, L. F., 'Descent of the Manor of Dixter', *SAC* 52 (1909).
— *English Life in the Middle Ages* (Oxford, 1926).
— 'Some Sussex Domesday Tenants, II: the Family of Dene', *SAC* 58 (1916).
— 'The Property of the Earl of Arundel, 1397', *SAC* 91 (1953).
— 'The Rapes of Sussex', *SAC* 72 (1931).
— 'Tregoz', *SAC* 93 (1955).
Saul, Nigel, *Knights and Esquires: the Gloucestershire Gentry in the Fourteenth Century* (Oxford, 1981).
— 'Murder and Justice, Medieval Style: the Pashley Case, 1327–8', *History Today*, 34 (Aug. 1984).
— 'The Despensers and the Downfall of Edward II', *EHR* xcix (1984).
Searle, E., *Lordship and Community: Battle Abbey and its Banlieu* (Toronto, 1974).
Simpson, W. D., 'The Moated Homestead, Church and Castle of Bodiam', *SAC* 72 (1931).
Somerville, R., *History of the Duchy of Lancaster*, i (London, 1953).
Stephenson, M., *A List of Monumental Brasses in Surrey* (reprinted Bath, 1970).
Stone, E., 'Profit-and-loss Accountancy at Norwich Cathedral Priory', *TRHS* 5th Ser. 12 (1962).
Streeten, A., *Bayham Abbey* (Sussex Arch. Soc. Monograph, 2, 1983).
Thiebaux, M., 'The Medieval Chase', *Speculum*, 42 (1967).

Thorold Rogers, J. E., *A History of Agriculture and Prices in England* (7 vols., Oxford, 1866–1902).
Thrupp, S., *The Merchant Class of Medieval London* (Chicago, 1948).
Tomkinson, A., 'Retinues at the Tournament of Dunstable', *EHR* lxxiv (1959).
Tout, T. F., *Chapters in the Administrative History of Medieval England* (6 vols., Manchester, 1920–33).
— *Collected Papers of Thomas Frederick Tout* (3 vols., Manchester, 1932–4).
Trevor-Roper, H. R., 'History and Imagination', in *History and Imagination: Essays in Honour of H. R. Trevor-Roper*, ed. H. Lloyd-Jones, V. Pearl, and B. Worden (London, 1981).
Trow-Smith, R., *A History of British Livestock Husbandry to 1700* (London, 1957).
Vale, M. G. A., *Piety, Charity and Literacy among the Yorkshire Gentry, 1370—1480* (Borthwick Papers, no. 50, 1976).
Victoria History of the Counties of England (in progress, London, 1900–).
Walker, S., 'Lancaster v. Dallingridge: A Franchisal Dispute in Fourteenth Century Sussex', *SAC* 121 (1983).
Ward, G., 'The Aldhams', *Arch. Cant.* xl (1928).
Ward, J., 'Fashions in Monastic Endowment', *Jnl. Eccles. Hist.* 32 (1981).
Wood S., *English Monasteries and their Patrons in the Thirteenth Century* (Oxford, 1955).
Wright, S., *The Derbyshire Gentry in the Fifteenth Century* (Derbyshire Rec. Soc. viii, 1983).
Wrottesley, G., *Pedigrees from the Plea Rolls*.

D. UNPUBLISHED THESES, REPORTS, AND OTHER WORKS

Bertram, J., 'Sussex Brasses and Slabs, Part I: West Sussex; Part II: East Sussex'.
Brent, J. (née Wooldridge), 'Alciston Manor in the Late Middle Ages' (Bristol MA thesis, 1965).
Clough, M., 'The Estates of the Pelham Family' (Cambridge Ph.D. thesis, 1956).
Homan, W. D., 'Winchelsea: the Founding of a Thirteenth-Century Town' (1940: typescript in ESRO).
Martin, D., 'An Architectural History of Northbridge and Salehurst Villages', in *Historic Buildings in Eastern Sussex*, 2, i (Rape of Hastings Architectural Survey, 1980).
Rogers, A., 'The Parliamentary Representation of Surrey and Sussex' (Nottingham MA thesis, 1957).
Titterton, J. E., 'The Malyns Family: Medieval Manorial Lords', dissertation submitted for the Certificate in Local History, Oxford, 1985.

INDEX

Members of families are listed chronologically and, where appropriate, in order of succession to the inheritance.

Alciston (Sussex) 119, 132
Aldham family 67
Aldington (Kent) 39
Aleyn, Geoffrey 19, 120 n.
  William 120
Alfriston church (Sussex) 143–4
Alveton, John de 183
Amberstone (Sussex) 9
Andrew, John 167
arbitration 20, 88–9
Arlington (Sussex) 12
Arnald, John 11
Arundel, Sir John 42
Ashburnham, John de 45, 65 n., 68
  Sir Roger de 45, 65 n., 68–9, 82, 139, 166
  Thomas, 69 n.
Ashdown, forest of 189, 191
Aubry, Bartholomew 11

Badlesmere, Bartholomew 11
Bainden (Sussex) 20, 21, 26, 39, 83, 88–9, 114, 120
Bainden, Henry 130, 134
  William 120
Barnaby, John 114
Basing, Margaret de 85–7
Battle, abbey of 43–6, 110, 141–2, 152–3; stewards of 45–6
Batsford, William de 3, 139
Bavent, Sir Roger de 181
Bayham, abbey of 154
  abbot of 84
Beche, Sir John de la 11, 12
  Sir Nicholas de la 12
Beddingham (Sussex) 3, 101, 109, 111–19, 121, 123, 125–9, 132, 133, 134, 135, 177 n.; account rolls of 92–3, 100, 105, 111–12, 125–9, 175
Belknap, Sir Robert 3 n., 45
Bergholt (Essex) 9, 10
Berkeley family 143
Bigod, Sir John 31, 32, 74
  Sir Ralph 31, 32
Bilsham (Sussex) 33
Blakehou, Roger de 11
Blast, Thomas 47
Boarzell (Sussex) 169
Bodiam (Sussex) 67, 68, 82, 166, 167
Bolney, Bartholomew 46
Bowley (Sussex) 9, 10
Boylyng, John 120
Bradford Bryan (Dorset) 3
Brambletye (Sussex) 166, 184 n.
Braose, Sir William de 34
Brenchley, Sir William 34
Brenzett (Kent) 3
Brien, Sir William de 7, 61
Brittany, John duke of 28
  John de Montfort, duke of 148–9
Brook, John 45, 47, 62
Brookland (Kent) 25
Browe, Sir Hugh 36
Bryanston (Dorset) 3, 64
Buckhurst (Sussex) 9, 10, 166
Bures Mount (Essex) 9, 10
Burgess, Joan *see* Sackville, Sir Andrew III
Burne, John de 179 n.
Bush, William 131
Buxhill, Sir Alan 25, 64, 142
  Thomas 64
  Sir Alan 64

Cade, William 114
Camoys, Sir Ralph 34 n.
  Sir Thomas 36, 47
Canterbury, abps of 38–41, 46
Chalvington (Sussex) 9, 11, 107, 110–11, 124, 129
chapels, domestic 157–60
Chichester (Sussex) 58, 176, 177
Chiddingly (Sussex) 12, 133
Claverham (Sussex) 12, 107, 124, 133–9
Clere, Robert 113, 114 n.
Cliffe (Sussex) 101, 176
clothing 171–3

Cobham, Sir Reginald 62
Lady Eleanor 171
Cooc, Gilbert 114
Cook, John 126, 128 n.
Crecy, battle of 17, 50, 183
Crioll, Sir John de 7 n., 149
crusading 174–5
Cudlowe (Sussex) 33

Daleham, Richard 131
Dallingridge, Sir Roger (d. *c*.1380) 36, 38, 56, 67–8, 185 n.
Sir Edward (d. 1393) 13, 29, 36, 56, 60, 67–8, 72, 75, 182, 191
Sir John (d. *c*.1407) 46 n., 70
Dauntsey, Sir John 36
Dawe, John 130
Dean, West (Sussex) 93 n., 120, 129, 133, 135
Debenham (Suffolk) 9, 10
Dene, Ela de 9
Despenser, Elizabeth 183
Sir Hugh the elder (d. 1326) 51–5
Sir Hugh the younger (d. 1326) 51–5
Sir Hugh (d. 1349) 50–1
Dixter, Great (Sussex) 6, 171 n.
Dolseley, Simon 183
Drew of Pevensey 4

Ely, Roger de 134
Emmington (Oxon) 9
Esshing, Thomas 140
Etchingham (Sussex), church of 140–9
manor of 3 n., 101, 169, 185 n.
Etchingham, Simon de (*c*.1150) 4
Sir Simon de (d. *c*.1244) 4, 142
Sir William de, III (d. 1294) 4, 51, 142, 170, 191 n.
Sir William de, IV (d. 1326) 3, 4, 25, 35, 61, 63–4, 84, 93, 94–6, 99, 100, 142, 154, 163, 175, 178, 179 n., 185; his income 3, 115–16
Eve, his wife 3, 4, 61
Sir Robert de (d. 1327) 4–5, 25, 51–5, 88, 168–9, 179
Petronilla, first wife of 167, 179
Simon de, rector of Herstmonceux 5, 25, 87–8, 94–7, 144–5, 170, 178
Sir James de (d. 1349) 5, 24, 103, 164, 170
Joan, wife of 170
John de, dean of S. Malling 5–6, 24, 102
Sir William de, V (d. 1389) 1, 3 n., 6–7, 24, 65 n., 81, 170 n., 176; his income 115–18, 125–8; his rebuilding of Etchingham church 140–1, 148–52, 155–60
Robert de, of Gt Dixter 6, 171
Sir William de, VI (d. 1412) 7 n., 61, 62, 83
Eu, counts of 4, 43, 142, 153
Ewhurst (Sussex) 69 n., 139

Fallesle, Sir John 36
'famuli' 119–25
Felton, Sir John 54
Ferrand, Bertrand 54
Fiennes, Sir William 69, 156
Sir Roger 72, 166
FitzAlan, Richard, earl of Arundel (d. 1376) 35, 49, 181, 182
Richard, earl of Arundel (d. 1397) 29, 67, 75; his retinue 36–7

Gaunt, John of, duke of Lancaster 28–9, 60, 70–2, 75, 148
Gentil, Sir Nicholas de 33
Glottenham (Sussex) 166–9, 179
Glynde (Sussex) 14, 20, 21, 39, 88–9, 110–11, 119–20, 122, 130–1, 132–3, 134, 180, 190–1
Godfrey of Malling 14, 39
Goring (Sussex) 119, 120
Grinstead, East (Sussex) 16

Hakelut, Sir Richard de 33
Halden, William 63, 184
Harengaud family 67; Sir John de 93
Harold, John 101–3, 113, 127 n., 176
Harvey, B. F. 131 n.
Harvey, P. D. A. 105 n., 125, 127 n.
Heighton St Clair (Sussex) 109, 129, 135
Hempsted (Kent) 3
Henry IV 70–1
Heron, Sir William 36
Herstmonceux (Sussex) 126 n., 156, 166 n.
Holland, John, earl of Huntingdon 69–70
Holmestede, William de 3 n., 101
Hoo, Peter de 63
Hoo, Sir Thomas 76, 103
households, size of 161–2
hunting 187–92
Hurstpierpoint (Sussex) 33
Hykeman, John 126

Ivychurch (Kent) 25

John of Gaunt *see* Gaunt, John of, duke of Lancaster

Kendal, Sir Edward de 183
Kenting, John 101 n.
'Knellesflote' (Sussex) 65, 164

Laughton (Sussex) 110, 133–4, 166, 192
Leem, John, prior of Michelham 89
Lewes, castle of 29–30, 35, 166, 176
 prior of 43, 191
 priory of 30, 35, 41–3, 176
 rape of 30, 59
Lewknor, Sir Roger de 178
Leye, John atte 128 n.
 William atte 126
Linch (Sussex) 3, 61
London 180–1, 183–6
Long, William 163
Lucas, John 163
Lullingstone (Kent) 3
Lymbergh, Adam 54
Lynde de la, family 67

Malemains, Sir Nicholas de 33
Malling, South (Sussex) 5, 38–9, 102
Malyns family 50 n., 61
Martin, David 166 n., 169 n.
Maufe, Sir John 11
Mayfield (Sussex) 38, 40
Medsted, Sir Philip 24, 36, 75, 191
Molyns, Sir John de 24, 49
Montagu, William, earl of Salisbury 49, 183
Mortimer, Sir Constantine 32
 Roger, earl of March (d. 1360) 17, 50
Mountfield (Sussex) 3, 94, 142
Mowbray, Thomas, earl of Nottingham 69, 70 n.

Neirford, John de 31
 Maud de 31
 Sir Thomas de 31
Netherfield (Sussex) 59
Newenden (Kent) 14
Newick (Sussex) 33
Northo, William de 34
Nott, John 183

Offington, Hamo of, abbot of Battle 43, 110
 Robert 114
Ore, William de (d. *ante* 1340) 65
 John de (d. 1361) 5, 24, 65
 Robert de (flor. 1370s) 1, 17, 65
Osberton, William de 183
Oxenbridge, Robert de 46, 47, 63
Oxford, earl of 191

Padbury (Bucks.) 3, 25
Pagham, William de 45
Parkyn, John 126
Pashley, manor of 86 n., 166 n.
Pashley, Sir Edmund de (d. 1327) 85–7, 184, 185
 Joan, wife of 86–7
 Edmund, son of 85–7
 John, son of 87, 91
 Sir Thomas de (d. *ante* 1361) 35
 Sir Edmund de (d. 1361) 185
 Sir Robert de (d. *ante* 1397) 1
Patching (Sussex) 19, 20, 21, 88–9
Paynel, Sir william de 33
Peakdean (Sussex) 3, 133, 135
Pelham, Sir John 46, 63, 70–2, 133–4, 139
Pelham, R. A. 111
Percy, Sir William 36
Pevensey, rape of 28–9, 59–60, 70, 71, 191
Peverel, Sir Andrew 38
 Lady Katherine 42
Pierpont, Sir Simon 33
Philip, William 126
Philippa, Queen 23–4, 29, 191
Pocock, Henry 120
Pole, Sir Edmund de la 13
Possingworth, John 113, 114 n.
Poynings church (Sussex) 143–4
Poynings, Sir Michael de (d. 1314) 33, 37
 Sir Thomas de (d. 1339) 37
 Sir Michael de (d. 1347) 17, 35 n., 37, 56
 Sir Michael de (d. 1369) 37 n., 67, 144, 149 n.
 Sir Thomas de (d. 1375) 36
  his widow 42
 Sir Richard de (d. 1387) 37 n., 129
Preston, John 41, 47
 Thomas de 63, 103
Pyecombe, Michael de 92–3, 100, 184

reeves, manorial 101, 125–9
Reinbert 4, 141
Richard II, King 69–70, 72
Robertsbridge, abbey of 4, 64, 94–7, 141–2, 144–5, 154, 164
Ros, Peter de 65–6

Rye (Sussex) 52, 70, 178
Rye, Henry of 44

Sackville, Sir Jordan de (d. 1273) 10
Sir Andrew de, I (d. *c.*1290) 10
Sir Andrew de, II (d. 1316) 11, 64
Sir Andrew de, III (d. 1369) 1, 48–51, 55, 63, 103, 171, 174–5, 182–3, 185, 191; his wardrobe 171–3
Joan, 1st wife of 13
Maud, 2nd wife of 13
Joan Burgess, mistress of 13, 27, 51
Andrew, son of (d. *ante* 1369) 13 n., 191
Sir Thomas (d. 1432) 13, 36, 46 n., 62, 75, 154 n., 191
St John, Sir Edward 42, 55, 83, 182
Sir John 36
Salehurst (Sussex) 3, 94–6, 142, 145, 146, 164
Savage, Sir Arnald (d. 1410) 7, 61, 63, 171
Sir Arnald (d. 1420) 7 n.
schooling, provision for 163
Scotney castle (Sussex) 45, 68, 166
Seaford (Sussex) 59, 132, 134 n., 177
Sedlescombe (Sussex) 31 n., 59
Sharnden, Robert de 45, 48, 190
sheep farming 132–6
Snave (Kent) 25
Socknersh (Sussex) 64
Sounde, John le 126
Peter 126
Robert 126, 128 n.
'Southall' (Sussex) 21
Spicer, John 63, 103
Stangrave, Sir Robert 34
Stanley, Sir John 72
stewards, in monastic employment 44–6; in secular employment 100–4; training of 102–3
Stopham (Sussex) 3, 61, 175
Stopham, Ralph de 3
Eve de *see* Etchingham, Sir William de, IV, wife of

Tarring, West (Sussex) 14, 39
Tauk, Sir William 1, 65 n., 184
Tawton, Robert de 95–6, 145
Thannington (Kent) 14, 21, 25, 39
Tiptoft, Sir Payn 36
Tregoz, family of 37, 64, 119, 186
trespass, action of 79–85
Trevor-Roper, H. R. 165
Turk, Sir Robert 180

Udimore (Sussex) 3, 6 n., 82, 94, 101, 118, 124, 136–8, 142, 170 n., 178 n.

Vienne, Sir Luke de 33
Sir Peter de 33

Waldern (Sussex) 12
Waleys, Sir Richard I (flor. *c.*1200) 14, 39
Sir Godfrey I (d. *ante* 1237) 14, 39
Sir Richard II 61
Sir Richard III 25
Sir Godfrey III (flor. *c.*1310) 16, 25, 39
Sir John I (d. 1376) 1, 16–18, 24, 39, 51, 56, 131, 134 n., 163 n., 166, 190–1
Andrew, son of 18, 174
Sir William I (d. *ante* 1409) 18–20, 25, 36, 56, 83, 88–9, 162, 174
Richard, half-brother of 20, 25, 83, 88
John II (d. 1418) 20, 83, 88–9, 180
Joan, wife of 20, 180
William, 'the idiot' 14, 89
Warbleton (Sussex) 168
Warde, Sir Roger le 183
Wardedieu family 67
Warenne, John de, earl of Surrey (d. 1347) 30–1, 73–4; his retinue 31–5
Joan, wife of 31, 67
Sir William de 35, 73–4
Welle, Peter atte 126–7
Thomas atte 128 n.
Whatlington (Sussex) 142
Whatlington, John of 44
Whebenham, William 120
Whitecliff, John 17
Wigsell, Great (Sussex) 169
Winchelsea (Sussex) 52, 58, 59, 118 n., 177–80
Winchester, John de 134
Wisham, Sir John de 32–3
Wolvercote, Sir William de 37 n.
Worth, Simon de 100

Yapton (Sussex) 3, 61